A Hand-Book to the Peak of Derbyshire

A HAND-BOOK

TO THE

PEAK OF DERBYSHIRE,

AND TO

THE USE OF THE BUXTON MINERAL WATERS.

South and West Fronts of the New Ranges of Hot Baths at Buxton, showing the connection of the Baths with the East End of the Crescent.

A HAND-BOOK

TO THE

PEAK OF DERBYSHIRE,

AND TO THE USE OF

THE BUXTON MINERAL WATERS;

OR,

BUXTON IN 1854.

By WILLIAM HENRY ROBERTSON, M.D.,

SENIOR PHYSICIAN TO THE BUXTON BATH CHARITY.

WITH

A MAP OF THE PEAK OF DERBYSHIRE, AND THE SURROUNDING DISTRICTS;
A PLAN OF THE BUXTON PARK AND PLEASURE-GROUNDS;
ELEVATIONS AND PLANS OF THE BATHS, ETC.;

A Botanical Appendix

By MISS HAWKINS;

AND

A DIRECTORY OF DISTANCES, ROUTES, HOTELS, INNS, AND LODGING-HOUSES.

LONDON:
BRADBURY & EVANS, 11, BOUVERIE STREET.
1854.

LONDON:
BRADBURY AND EVANS, PRINTERS, WHITEFRIARS.

PREFACE.

THE want of such a work as the present has been so continually and urgently complained of, that I hope this volume may prove to be acceptable to the public. I have made occasional use of my work on "Buxton and its Waters." The present hand-book is, however, of a much more extensive character; at the same time that it is not intended to answer all the more strictly medical purposes of the older publication. I am so much indebted to Miss Hawkins, for the excellent botanical commentary and catalogue,—to Mr. Smithers, the able and indefatigable agent of his Grace the Duke of Devonshire, for valuable suggestions, and kind and efficient assistance in every way,—to Mr. Henry Currey, the architect of the new baths and principal buildings, for permission to use the elevations and plans,—and to Mr. Smith, of the Buxton office, for the excellent map of the district, and plan of the new park, walks, and pleasure grounds,—that it is my grateful duty to record the obligations, and express my thanks.

THE SQUARE, BUXTON,
June, 1854.

CONTENTS.

CHAPTER I.

HISTORICAL RECORDS OF BUXTON AND ITS BATHS.

CHAPTER II.

PHYSICAL CHARACTER AND ITINERARY OF BUXTON AND THE PEAK OF
DERBYSHIRE.

CHAPTER III.

ORIGIN AND CAUSE OF THE HEAT OF THE THERMAL SPRINGS OF
BUXTON.

CHAPTER IV.

GENERAL PROPERTIES OF THE BUXTON TEPID WATERS; THE SUC-
CESSIVE ANALYSES, AND THEIR RESULTS; AND COMMENTARY
ON THEIR COMPOSITION, IN REFERENCE TO THEIR MEDICINAL
EFFECTS.

CHAPTER V.

ASCERTAINED ELEVATIONS OF DIFFERENT PARTS OF THE DISTRICT.—
THE NEW RANGES OF BATHS.—AMOUNT OF FLOW OF THE TEPID
SPRINGS.—THE WELLS, BATHS, AND DOUCHES.

CHAPTER VI.

PRIMARY, SECONDARY, AND ALTERATIVE EFFECTS OF THE BUXTON
TEPID WATERS.—MORBID CONDITIONS FOR THE RELIEF OF WHICH
THEY ARE USEFUL.—CIRCUMSTANCES WHICH CONTRA-INDICATE
THEIR USE.—RULES FOR THE USE OF THE BATHS, AND FOR
DRINKING THE TEPID WATERS.

CHAPTER VII.

ANALYSIS, CHARACTER, AND USES OF THE CHALYBEATE WATER
OF BUXTON.

CHAPTER VIII.

THE GRITSTONE WATER.—AMOUNT AND CHARACTER OF ITS SUPPLY
FOR DOMESTIC AND ORDINARY PURPOSES AT BUXTON.

CHAPTER IX.

HISTORY, PROGRESS, POSITION, AND USEFULNESS OF THE BUXTON
BATH CHARITY.

APPENDIX.

DIRECTORY, &c.

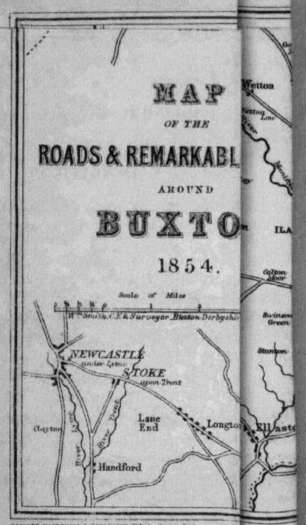

MAP

OF THE

ROADS & REMARKABL

AROUND

BUXTO

1854.

Scale of Miles

Wᵐ Smith, C.E. & Surveyor, Buxton Derbyshi

NEWCASTLE
under Lyme

STOKE
upon Trent

Wetton

ILA

Calton
Moor

Swinsco
Green

Stanton

Clayton

Lane
End

Longto

Ellasto

Handford

MACLURE, MACDONALD & MACGREGOR, Lithographers to the Queen.

A HAND-BOOK

TO

THE PEAK OF DERBYSHIRE,

AND TO THE USE OF THE BUXTON MINERAL WATERS.

CHAPTER I.

HISTORICAL RECORDS OF BUXTON AND ITS BATHS.

"Buxton is situated on the western side of the north part of the county of Derby, in a tract of elevated, uneven, and hilly *moorland*, called, therefore, the *High Peake*, or the *Peake Hundred*.

"The Peake is about fourteen or sixteen miles broad from the south-west to the north-east side; its whole length, from the north-west to the south-east, may be twenty miles; and it is supposed to contain one-fourth part of the whole county, or 170,000 acres.

"This region of high land is the southern extremity of the ridge or chain of mountains and hills, that extends from the Cheviot hills in Scotland, nearly through the middle of the island, and terminates in the north part of Derbyshire.

B

As this range of eminent land runs through the middle of the north of England, as the Apennine does through Italy, it has been called the *English Apennine*.

"The British Apennine may be reckoned, for the sake of forming a general conception of it, from fifteen to twenty miles broad. Near Scotland it is much broader, and as the island to the north of Derbyshire contracts itself considerably in breadth, this tract of high land bears no small proportion to the breadth of the north part of England.

"The whole length of this ridge of land appears to be about a hundred and forty miles."

The above is a quotation from a large work, in two octavo volumes, by Dr. Pearson, an eminent physician and chemist of the time, and published in 1784. But the upland country on which Buxton is situated, must have much larger dimensions assigned to it. From Ashbourne on the south, at a distance of twenty miles, to beyond Glossop on the north, Buxton being placed almost centrally in reference to these boundaries, the hill country rises from the lower districts of Derbyshire and Staffordshire on the one side, and of Lancashire and Cheshire on the other. Arising again from Staffordshire and Cheshire on the west, to the east of the towns of Leek, Congleton, and Macclesfield, at a distance of about twelve miles from Buxton, the upland country extends eastward into Yorkshire, stretching at length into Scotland. The upland district, which is virtually the Peak district, and almost in the centre of which Buxton is placed, may be said to have a diameter of between thirty and forty miles, in all directions. Presenting a varying elevation, from a few hundreds to nearly two thousand feet higher than the level of the sea,—and nearly the whole of this great surface of country being divided, and almost in equal proportions, between the formation of millstone grit and that of secondary limestone,—the scenery is characterised by large outlines, massive boldness, and great variety ; the mountain masses, sloping hill-sides, broad basin-like valleys, and moorland summits of the gritstone, contrasting with and varying the more abrupt and fantastic grandeur, the summits of bare and rugged rock, the sharp outlines, and the narrow and rocky valleys of the mountain limestone.   Overlooking the lower districts around it, in all directions ; and offering numberless pictures of more confined character in its own valleys, shut in by its own hills,—many of these scenes, however, having an extent of many miles ; this great upland region is deservedly considered to be one of the most picturesque and beautiful districts in Great Britain.

According to the statement of Messrs. Lysons, in their History of Derbyshire, the word Buxton was written Bawkestanes, in the time of King Henry the Third ; and they add, " it seems probable that it was originally written *Badestanes*, deriving the name from its stone baths."

Dr. Short, in his " History of Mineral Waters," presented and dedicated to the Royal Society, and published in 1734, states that, " without all dispute," the Buxton baths must have been well known to the Romans.   It seems that, in 1709, Sir Thomas Delves, of Cheshire, in memory of

having been cured by these waters, caused an arch to be
erected over one of the springs, "twelve feet long and
twelve feet broad, set round with stone seats on the
inside;" and "in the middle of this dome, the water
sprung up in a stone basin, two feet square above." In
preparing the site for this erection, which in its turn had to
be removed, when extensive buildings were erected in
1780—1784, "an ancient Roman brick wall about St
Anne's well" had to be removed. "In 1698, when Mr.
White, then of Buxton Hall, was driving up a level to the
Bath, fifty yards east of St. Anne's well, and fourteen yards
north of Bingham spring, the workmen found, buried under
the grass and corn-mould, sheets of lead, spread upon great
beams of timber, about four yards square, with broken
ledges round about, which had been a leaden cistern, and
not unlikely that of the Romans, or some other ancient
bath, which had been supplied with water from Bingham
well. Thirdly, the Roman highway from Burgh (Brough), a
small village twelve miles east, to Buxton, a great part
whereof remains entire to this day, reaches within half
a mile of Buxton-Hall; and not improbably it took a turn
from Burgh to Castleton, two miles north-west; for above
this, on the top of Maniton, is remaining a very beautiful
and strong camp. All for two miles below, is a fortified
station four square, the town a garrison, and the castle
above it a fort, armoury, or watch-tower, to answer the
camp. Fourthly, that it was of great repute in the darkest
distant times is undeniable, from the chapel here dedicated

to St. Anne, whose foundation was likewise discovered, and a large piece of its wall dug up in driving the aforesaid level."

Dr. Leigh, who seems to have died about 1671, and who was one of the numerous writers on the subject of the Buxton waters, says, that, in his time, a wall was to be seen near St. Anne's well, which he believes to have been of Roman erection. He describes it as cemented with plaster, red and hard as brick, but very different from anything at that time in common use, having more the resemblance of some kind of tile than of any other substance. The ruins of an ancient bath, too, he says, were then visible, composed of matters similar to the wall, and so perfect in every part as to present to an observer every one of its dimensions. Mr. Pilkington, in a work published in 1781, observes, that "when the foundations of the crescent were dug, the shape and dimensions of this bath (speaking of one mentioned in Bishop Gibson's edition of Camden's Britannia, as visible near St. Anne's well) might be very easily discerned. Its form was that of an oblong square: it measured thirty feet from east to west, and fifteen feet from north to south. The spring was at the west end of the bath; and at the east end there had evidently been a floodgate for letting out the water. The wall was built of limestone, and appeared to be of rude workmanship. On the outside, it was covered with a strong cement; supposed to have been for the purpose of preventing cold water from mixing with the warm spring supplying the bath. The floor was formed of plaster, and appeared to have been uninjured by time. On the top of the walls were laid strong oak beams, which were firmly connected together at the four corners; and the bath had the appearance of having been exposed to the air."

Dr. Jones, who published a work entitled "The Benefit of the Ancient Bathes of Buckstones," in 1572, says, "Joyning to the chief spring, between the river and the bath, is a very goodly house, four-square, four stories high, so well compact with houses and offices underneath, and above, and round about, with a great chamber, and other goodly lodgings to the number of thirty, that it is and will be a beauty to behold, and very notable for the right honourable and worshipful that shall need to repair thither, as also for others, yea, and the poor shall have lodgings and beds hard by for their uses only. The baths also are bravely beautified with seats round about, and defended from the ambient air, and chimneys for fire to air your garments, in the bath side, and other necessaries most decent. And truly, I suppose, that if it were for the sick a sanctuary during their abode there, for all causes saving sacrilege, treason, murther, burglary, and robbing by the highway-side, with also a license for the sick to eat flesh at all times, and a Friday market weekly, and two fairs yearly, it should be to the posterities not only commodious, but also to the Prince great renown and gain."

Such are some, out of many, of the curious and not un-interesting accounts of Buxton, in its more ancient days.

There seems to be every probability, that at least two of the great ancient roads met at Buxton. One of these has already been noticed, in the quotation from Dr. Short's important work; the part of which between Brough, or Burgh, a Roman station near Hope, and Buxton, was traced by Mr. Pegge, in the year 1779. This road seems to have extended from Middlewich, Congleton, Buxton, and Brough, to York and Aldborough. The part between Buxton and Brough is still called Batham-gate. Another of these great roads, extended from Manchester to Buxton, and thence southward, under the names, in different parts, of High Street, Street-Fields, Street-Lane, Old-gate, &c. The parts of this road which are still noticeable, extend from Bollington, about thirteen miles from Buxton, cross the higher grounds by Pym's Chair, and descend thence to the valley of the Goyt, being continued as far as Goyt's bridge, within three miles from Buxton. The road may have been continued up the valley, by the side of the river Goyt, to Goyt's Clough; or, more probably, was carried across the river, and up the opposite hill-side, near to, or on the site of, the existing Goyt's lane. Immediately to the south of Buxton, this road is again noticeable, near to Coteheath, close to the high road to Ashbourne,—and again, about five miles from Buxton, near to the Duke of York, public-house, on the left hand side of the same road. Whether these roads were originally constructed exclusively for military purposes,—for effecting the conquest, or more complete subjugation, of the people of the country; or whether they may have only been the means

of communication between important places; it seems to be
evident that they would bring into notice, if they were not
constructed for the convenience of, the places through which
they might pass; and the Buxton waters, with their
elevated temperature, large flow, and medicinal value,
would acquire repute at very remote periods of time.   That,
at periods so remote from our own, large baths should have
been constructed, of such a durable and costly character as
a frame-work of wooden beams lined with lead, in one
instance,—and, in another, of masonry, floored with concrete,
and most carefully protected outside with thick and strong
cement (even although these structures may have been
uncovered and exposed to the open air), cannot fail to
astonish us.   Now that modern roads and modern railways
have served to bring distant places so much more within
reach of one another, the Peak of Derbyshire is still
sometimes thought to be too remote from the southern parts
of England; and it may well be thought wonderful, that
sufficient numbers of people should have made the pilgrimage to Buxton, in those early times, in search of health
from the use of its waters, to have led to the formation of
such baths as these.   Even now, in the baths which have
just been re-constructed, and so much extended, the largest
of them is twenty-six feet long and eighteen feet wide;
and we find the measurement of one of these old baths,
probably the work of the Romans, to have been thirty-feet
long by fifteen feet wide.   The bath of lead and oak seems
to have been of older date than this, and at least twelve feet

square; and either will tell a tale full of import, as to the
enterprising spirit of a time so distant from our own, as to
the estimation in which the Buxton waters were then held,
and as to the still more ancient period at which they must
have begun to be famous for their medicinal qualities.

In the midst of all this activity of hand, and skilled
labour, there was no printing-press, and there were few
scribes; the affairs of remote provinces had no historians;
traditions and public report were the histories and news-
papers of the age; and but for accidental discoveries of
such magnitude, the ancient use of the Buxton waters, and
their ancient fame, would have been unknown to us. And,
even down to the age of Queen Elizabeth, there is no
history of Buxton, nor account of its waters; although the
reputation of their curative efficacy had become so considerable, that the accommodations of the place were no
longer equal to the wants and demands of the people
resorting to it; and a large building was in consequence
erected by the Earl of Shrewsbury, at that time the
principal proprietor of Buxton and the estates adjoining.
This building, so quaintly described by Dr. Jones in the
quotation already given, must have been regarded as a fine
and imposing edifice, even in that less ancient time. It
supplied accommodation to some of the principal per-
sonages of that age, who visited Buxton for the use of its
waters.

Mary, Queen of Scotland, visited Buxton at least four
different times, while in the custody of the Earl of

Shrewsbury. These visits must have occurred between
the years 1570 and 1583, inclusive. There is a curious
account of the circumstances leading to and connected
with these visits, in Lodge's "Illustrations of British
History." From this authority we likewise learn, that
the Buxton waters were used for the relief of their ailments, by two of the greatest men of those times, viz, the
Earl of Leicester and Lord Burleigh. Miss Strickland's
edition of the "Letters of Mary Queen of Scots, and
Documents connected with her Personal History," may
also be referred to. One of the visits of the great Lord
Burleigh is noticed in one of these letters. This visit took
place at the time when Queen Elizabeth was sojourning at
Kenilworth. In a letter, dated, Buxton, August 10th,
1579, the Queen mentions the benefit which she had
derived from the use of the baths, in relieving a severe pain
of the side. The Queen's last visit to Buxton seems to
have been in the year 1583. When the statements contained in these letters are considered, as to the condition of
the places in which the poor Queen was confined,—the
extremely damp state of the grounds, and buildings, and
even of the apartments at Tutbury, which this royal lady
was made to occupy, affording even an inadequate shelter
from the weather,—it seems to be probable that her
ailments were of rheumatic character; and for the relief of
such, the use of the baths and waters of Buxton would
be of no mean value.

The Queen is said to have scratched on a pane of glass,

in a window of the room she occupied, the following classical and kindly farewell :—

Buxtona, quæ calidæ celebrabere nomine lymphæ,
Forte mihi posthac non adeunda, vale !

It is stated, in Camden's Britannia, that this distich is an adaptation to Buxton, of Cæsar's verses upon Feltria. The relief afforded to Queen Mary's case, appears to have induced both Lord Burleigh and the Duke of Sussex to resort to Buxton, for the cure of their ailments. As Lord Burleigh visited Buxton at least twice, viz., in the years 1577 and 1580, it seems to be a justifiable inference, that the baths were of use to him. The visit of the Duke of Sussex is mentioned as having occurred in July, 1580.

It is difficult to realise the various circumstances and position of times so remote as even those of Queen Elizabeth, and to picture the condition of Buxton, and of the inhabitants of the town and the adjoining hamlets, in those days. There is, in the Chapel of the Rolls, the original record of " A Grant to Thomas Dakyn and the inhabitants of the Chapelry of Fairfield," dated the thirty-seventh of Queen Elizabeth (A. D. 1595), which illustrates curiously the state of Buxton at that time, and contrasts very much with what obtains at present. Fairfield is a pretty village and chapelry adjoining to Buxton ; and, it is, in the present day, much indebted to its close proximity to Buxton for an enhanced value of its land, and as affording a ready sale for its agricultural produce ; the village, moreover, being advantaged by affording lodgings to some of the visitors of Buxton. In those days, however, the people of Fairfield appear to have suffered more from the poor frequenters of Buxton, than they gained, either directly or indirectly, from being so near to a place of such resort; and, accordingly, the inhabitants humbly supplicated the Queen for a grant to support a " minister or chaplain," pleading in the supplication, among other weighty reasons justifying the royal bounty, that " the Inhabitants aforesaid had fallen into extreme poverty," stating that the said poverty was in part " by reason of the frequent access of divers poor, sick, and impotent persons repairing to the Fountain of Buxton, in the county aforesaid, within the neighbourhood of the Chapel aforesaid, for whose maintenance and relief the Inhabitants aforesaid are daily charitably moved to apply their own goods, by which the aforesaid inhabitants of the Chapelry aforesaid are not only rendered unable to sustain and maintain the Minister or Chaplain aforesaid, but also, by reason of their poverty, the aforesaid Chapel has fallen into great ruin and decay, and thus the inhabitants aforesaid will be altogether deprived of all Divine Service and Spiritual Instruction, unless a speedy remedy, in this behalf, shall be provided by us, wherefore they have humbly supplicated us, that we (being piously inclined) should be pleased to found and establish, within the town of Fairfield aforesaid, one perpetual Chapel, for our minister or chaplain to celebrate divine service there, for all the inhabitants, within the chapelry of Fairfield aforesaid, for ever to remain." I am indebted for this

interesting reference to the kindness of the late Mr.
Goodwin, of Pigtor; and it furnishes a curious picture of
the times; illustrating the difficulties with which poverty
and disease must have had to contend in so much greater a
degree, when the means of travelling from remote distances
must have been most tedious and expensive, and when the
journey of a poor sick person to Buxton must indeed have
been a difficult and severe undertaking. It shows, however,
how much the use of these waters must have been valued
even at that remote period, when such difficulties and severe
privations had not so checked the visits or kept down the
numbers of these poor seekers after health, but that they
should have proved to be so great a tax and burthen to the
inhabitants of surrounding hamlets. The poorest can now
find means of transport; and the visit to Buxton is never,
in these days, the weary and trying pilgrimage which it
must have been to poor sufferers in the days of Queen
Elizabeth. Two years after the grant above-mentioned, in
the thirty-ninth year of the Queen's reign, perhaps in
consequence of the supplication from the people of Fairfield,
perhaps from increasing resort to Buxton, and further
supplications, it was enacted, "that none resorting to Bath
or Buxton Wells should beg, but should have relief from
their Parishes, and a pass under the hands of two Justices of
the Peace, fixing the time of their return, nor were they to
beg there under pain of incurring the penalties of that act."

Previous to the period of the Reformation, the medicinal
effects of the Buxton waters had been ascribed to the

saintly influence of their great patroness, St. Anne; and
the walls of a chapel that was dedicated to her, had been
decorated, from time immemorial, with the crutches of those
cripples who had been cured by the use of these baths, and
who no longer required them. In the earlier years of the
reformed religion, Buxton was made to suffer on account of
the superstitious errors of its earlier patrons. Conceived to
aid in keeping up a belief in the Romish doctrine of saintly
interference in human affairs, these interesting memorials
of gratitude for restored health were destroyed; and indeed
so bigoted had the national feeling, or rather perhaps the
feeling of the dominant party, become, against everything
connected with the unpopular faith, that the waters were
for a short time prevented from being used, by public
authority. The following document in regard to this, is too
curious to be omitted. It is addressed to Lord Cromwell
by one of the agents employed by him, for the suppression
of all establishments connected with the Romish faith:—

"RIGHT HONOURABLE AND MY INESPECIAL GOOD LORD,

"According to my bounden duty, and the tenor of your
Lordship's letters lately to me directed, I have sent your
Lordship by this bearer, my brother, Francis Bassett, the
images of St. Anne of Buckston, and Saint Andrew of
Burton-upon-Trent; which images I did take from the
places where they did stand, and brought them to my house
within forty-eight hours after the contemplation of your said

Lordship's letters, in as sober a manner as my little and rude will would serve me. And for that there should be no more idolatry and superstition there used, I did not only deface the tabernacles and places where they did stand, but also did take away crutches, shirts, and shifts, with wax offered, being things that allure and intice the ignorant to the said offering; also giving the keepers of both places orders that no more offerings should be made in those places till the King's pleasure and your Lordship's be further known in that behalf.

" My Lord, I have locked up and sealed the baths and wells of Buckston, that none shall enter to wash there till your Lordship's pleasure be further known. Whereof I beseech your good Lordship that I may be ascertained again at your pleasure, and I shall not fail to execute your Lordship's commandments to the utmost of my little wit and power. And, my Lord, as touching the opinion of the people, and the fond trust they did put in those images and the vanity of the things, this bearer can tell your Lordship better at large than I can write; for he was with me at the doing of all this, and in all places, as knoweth good Jesus, whom ever have your Lordship in his precious keeping.

" Written at Langley, with the rude and simple hand of your assured and faithful orator, and as one and ever at your commandment, next unto the King's, to the uttermost of his little power.

" WILLIAM BASSETT, KNIGHT."

" *To* LORD CROMWELL."

The shutting up of the baths and wells would not appear to have been long enforced, nor the reputation of the waters to have been much influenced by these arbitrary and prejudiced proceedings. In truth, the cures which had been effected, during so many ages, could not be so set aside and ignored; and, as they were no longer to be considered as being attributable to saintly influence, they began to be ascribed to the properties of the waters themselves.

The Hall, " at that time reckoned a fine mansion,"—" a very goodly house, four-square, four stories high,"—and which appears to have been well frequented, was destroyed about the year 1670, " and a new edifice erected on its site by William, the third Earl of Devonshire." This mansion, with many alterations and very considerable additions, is still a principal hotel in Buxton, and still called the Hall. Speaking of this building and its surroundings, Dr. Short, writing sixty years after the time of its erection, says,—" Buxton Hall is situated on the south brink of the rivulet *Wye* or *We*, from the union of three springs, a short mile west of the house, called *I*, *Thou*, *He*, which being united obtain the plural *We*. On the north side of the river is a steep mountain, covered chiefly with heath, under which is a black moss or peat earth, below that a shale, then clay and coal, and lead in some places. The surface here is very barren, and therefore return we to the south side, which, for about two miles, is a mountain of an easy ascent. The ground all

about the warm springs, on the south side of the river, is
very dry, fruitful, and pleasant, being a thin, warm, fertile
mould, lying upon limestone ; the grass, though short, is
very sweet and fattening, hence they have the most delicious
beef and mutton. Snow lies a much shorter time here than
in the lower country.(?) Here is good store of hares and
foxes, several wild rabbits of the rocks, partridges, moor
game of two sorts, one a large black-cock weighing five
pounds a piece, the other a brown and much less, tho' more
plentiful. The small river which runs from west to east
abounds with fine trout, grellin (grayling), crayfish, and
silver eels. A little east of St. Anne's well, over the ditch or
level which carries the warm water from the bath, is made a
curious natural hot-bed, and upon the rest of this canal
might be made the finest greenhouse in the northern
kingdoms ; Mr. Taylor, of the Hall, has also taken in several
new gardens with planting, and several curious walks. The
garden-stuff has a peculiar, grateful flavour. Up one pair of
stairs in the Hall is a beautiful dining room, seventeen yards
long, and nineteen feet wide, seven other entertaining
rooms, eleven lodging rooms with single beds and closets,
twenty-nine other lodging rooms ; this one house affords sixty
beds for gentlemen and ladies, besides suitable accommodations for their servants, and all other proper or useful offices."

In the front of the Hall was " a pleasant warm bowling-
green, planted about with large sycamore trees;" and on the
north side of the green was a grove of trees, which extended
on the north side of the Hall, and on the south bank of the

river, sheltering the bowling-green, the Hall itself, and the wells and baths, from the northerly winds. St. Anne's well was situated on the east of the great bath, and very near to it; as nearly as might be, on the spot where the new St. Anne's drinking well has now been erected; and therefore some yards to the west of the well recently removed, which was situated at the foot of the terrace walks of St. Anne's cliff, opposite the Crescent. Close to the river and the grove of stately trees, at the back of the Hall, probably near to the site of the western end of the Crescent, were the gardens, which appear to have been at one time so well managed and productive; and beyond, and to the south and south-east of these gardens, the valley was divided into closes or small fields, in which the different wells were situated. In a work published in the year 1646, entitled "A Prospect of the most famous parts of the world," under the head " Darbyshire," is the following : —"Things of stranger note are the hot-water springs, bursting forth of the ground at Buxton, where out of the rocke, within the compasse of eight yards, nine springs arise, eight of them warm, but the ninth very cold." The street called the Spring-Gardens, evidently obtains its name from the gardens of the Hall, which were so famous in the time of Dr. Short.

Mr. Macaulay, in the first volume of his "History of England," (page 345), says :—"England, however, was not, in the seventeenth century, destitute of watering places. The gentry of Derbyshire and of the neighbouring counties repaired to Buxton, where they were crowded into low wooden sheds, and regaled with oatcake, and with a viand which the hosts called mutton, but which the guests strongly suspected to be dog." Mr. Macaulay gives as his authority for this statement, a "Tour in Derbyshire, by Thomas Browne, son of Sir Thomas." It has been seen, however, that, from a much earlier time than that mentioned by Mr. Macaulay, Buxton was not only a watering-place of much importance and resort, but that its principal hotel was a large and commodious house, supplied with all the comforts and requirements that were then to be obtained anywhere; and indeed, that during at least three centuries before the period at which Buxton is thus stigmatised, the wants and expectations of the public had been provided for in the fullest manner, by an amount and excellence of house-accommodation, and bathing-accommodation, that must have been considerably in advance of most other places of the same kind. The excellence of the mutton, so vaunted by Dr. Short, and so well-known in our own times, gives to the stigma a still more marked ironical character.

In the year 1780, according to Mr. Bray, the foundations of the great pile of building were laid, called from its form the Crescent; the architect having been the celebrated Mr. Carr, of York. This beautiful structure, which was finished in 1784, is still the finest crescent-shaped elevation in England, and probably in Europe; and affords extensive accommodation to a large number of visitors.

By the erection of this building, all the immediate

localities of the river, baths, wells, roads, &c., were
much altered.   The high-road from Manchester, which
seems to have passed near to the Hall previously, was
turned, and made to pass at the back of the large new pile
of buildings.   The greater part of the grove or avenue of
trees was cut down; those only being left which surrounded
the bowling-green of the Hall, and protected this piece of
ground on the north, most of which probably still remain.
The river was enarched the whole way from the Hall to
some distance beyond the eastern end of the Crescent; and
the space occupied by this arch, by the large part of the
avenue of trees that had been cut down, and by some of the
springs which had emerged near to the south bank of the
river, and by the closes of land on the river side, between
the river and a rocky mound called St. Anne's Cliff, was
occupied by new buildings, forming the Crescent and the
Square.   And, in course of time, the rocky bank, or
rounded and considerable eminence, fronting the Crescent,
and said to have been a most unsightly-looking foreground
to so palatial a structure, was forced into form and usefulness by the taste and skill of Sir Jeffery Wyatville, and
formed into ranges of terrace walks, with intervening grass
banks, adorned with vases, of form, and style, and size, to
correspond with the Crescent; the whole being made into a
foreground of pleasing and ornamental character.

Gradually,—in addition to the piles of building formed by
the Crescent, the Hall, the Square,—a row of houses on the
west side of St. Anne's Cliff, called the Hall Bank,—an inn,

the George,—and another inn, the Grove; in addition to a church, built at much cost by the Duke of Devonshire, the noble owner of the baths and adjacent property; in addition to a large range of building, erected on the rising ground on the north, and at the back of the Crescent, for stables and coach-houses, &c.; a street came to be formed on the south bank of the river, beyond the enarchment of the stream which is covered by the Crescent. All these buildings, however, with no exception of any importance but the Hall, are comparatively modern. Buxton, more strictly so called, distinguished now-a-days from the part of the town above-mentioned, which is called Lower Buxton, by being called Upper Buxton,—this, the old town of Buxton, is on a level with the summit of St. Anne's Cliff, and has an elevation of upwards of seventy feet above the lower and more modern part of the town. Higher Buxton, or Upper Buxton, contains a much older and smaller church, dissenting chapels, a spacious market-place, the Eagle inn (a large house which has been long in repute), and many smaller inns, and a great number of lodging-houses.

But for a long time after the Crescent had been built, and after many other additions and improvements had been made, to meet the wants of those resorting to it, chiefly for the use of the baths, Buxton had to contend with many local disadvantages. The town had been situated in the midst of a bare and barren tract of country; there was hardly a tree within miles of it, unless at the bottoms

of the more important valleys; the land was, for the most
part, unenclosed, uncultivated, and unsheltered from the
winds; and the whole district must have looked wild,
dreary, and inhospitable. Even within the memory of old
inhabitants, there was neither cultivation nor enclosure
within twelve miles, in the direction towards Ashbourne,
unless in rare and isolated patches; and nearly the whole
of the valley of Buxton, on the south-west and west of the
town, and within a stone's throw of the old bowling-green
and new church, was untouched moorland. And yet,
induced by the great and well-deserved reputation of its
healing waters, the invalided from all parts had been
content to visit and sojourn in this region of wild and
barren, if picturesque and mountainous beauty; and so
great were the benefits derived, that, without most of the
usual supplementary watering-place attractions, the Buxton
waters supported and added to their celebrity.

At length it was found, between forty and fifty years
ago, that, in this mountainous and large featured district,—
which, in the ancient times, had been well timbered, and
formed part of the great midland forest of England,—trees
would grow; if they were planted. It had been thought, notwithstanding the fine old hawthorn trees to be seen placed
here and there, in all sorts of situations, elevations, and
exposures, in different parts of the district, that the hawthorn would not thrive in the locality; and therefore that
hedgerows could not be substituted for stone walls, as
fences for the fields. Many hundreds of acres have been

planted from that time to this; and accordingly, although
such a country as this ought always to be characterised by
the bold and massive grandeur of its scenery, it no longer
conveys a sense of bleak desolation, which it must have
done half a century ago; and the country around Buxton is
now universally allowed to be beautiful. That Buxton
should have been yearly resorted to, by thousands of
invalids, under such disadvantageous circumstances, may be
accepted in confirmation of the power of its waters, in
relieving and curing disease.

But, however great and praiseworthy the efforts which had
been made, to render the town and district more and more
worthy of the resort of the invalided, much had still been
left undone. A larger and larger amount of house-accommodation had indeed been afforded from time to time, but still
much below the wants of the public; additional baths had
been made at distant intervals, but even this essential
requirement was not adequately provided for; public walks
and pleasure-grounds had been laid out and planted, with
a princely liberality, for the use of the inhabitants and
visitors; but the march of taste and science, in regard to
embellishment and drainage, had to be kept pace with. Such
extensions, additions, and improvements have now been
made; and with a success that is universally admitted to
have been entire. The state of Buxton now, with its noble
and extensive ranges of baths, supplied with all the acces-
sories which art and ingenuity and science can furnish,—
with its many and various pleasure-grounds, and promenades,

and plantation-walks, and ornamental shrubberies, some of
them being immediately contiguous to the principal buildings, all within easy access, and all thrown open freely and
gratuitously to the public,—with its park of more than a
hundred acres, laid out and planted for ornamental and
building ground, from plans by Sir Joseph Paxton,—with
its surrounding hills, clothed with plantations of thriving
trees, wherever plantations are desirable, either for the
purpose of shelter, or of beauty,—with its dry soil, and
tempered mountain air, and mountain climate,—this, the
Buxton of the year 1854, ought to be thus compared
with the place as it was, even thirty or forty years ago,
with the place as it was in the year 1838, when I first
published an account of Buxton and its waters, and even
with Buxton as it was only three years ago. To supply
and chronicle the materials for such a comparison, has been
a grateful motive for collecting the above-cited contribu-
tions towards a historical account of the Buxton waters;
and the history must be admitted to be one of progress,
and the future to be full of promise. Much as has been
done from time to time, for the improvement of Buxton,
and the advantage of its visitors, the increasing reputation
of the mineral waters, and the increasing resort of the
invalided, have always preceded and justified the extensions
and improvements. It is undeniable, that the character
and success of this important watering-place have been
singularly independent of any circumstances but the marvellous efficacy of its healing waters.

South Front of the New Ranges of Natural Baths at Buxton, showing the connection of the Baths with the West End of the Crescent.

CHAPTER II.

PHYSICAL CHARACTER AND ITINERARY OF BUXTON AND THE PEAK OF DERBYSHIRE.

THE lowest part of the town of Buxton is at an elevation of one thousand feet above the level of the sea. It is, however, surrounded on all its sides by hills of greater elevation; and it occupies the north-eastern extremity of an oblong basin, the bottom of which is between two and three miles long, and about half a mile in breadth. The surrounding hills rise from the bottom of the valley by shelving sides, which gives to the upper margin of the basin a diameter of from four to eight miles, in different directions. The hills which bound the valley of Buxton, rise from it with very different degrees of abruptness. On the north and north-west, within little more than a mile from the town, to the right of the road to Manchester, Black Edge, the highest part of Comb's Moss, has an elevation of 1670 feet. On the west, at the distance of between two and three miles, and to the right of the Leek road, and to the left of the roads which branch from this road to Congleton and Macclesfield, the highest part of the chain of hills has an elevation of

o

nearly 2000 feet. This is a well known and commanding
ridge, called Axe Edge. On the south, the highest part of a
chain of hilly grounds has an elevation of 1435 feet. These are
now covered with what are known as the Grin Plantations,
and were formerly, and at their more distant extremity
are still, the site of extensive lime-kilns. The nearest part
of this range of high grounds is within less than a mile from
the town. On the south-east, Chelmorton Low forms the
highest part of the range of hills. This Low, probably one
of the many stations for signal fires in ancient times in
these upland districts, has an elevation of 1474 feet.
Chelmorton Low is at a distance of five miles from the
town, to the left of the road to Ashbourne, and to the right
of the road to Bakewell. But between Chelmorton Low
and Buxton, there is a considerable elevation of high land,
called Stadon. Almost due east from Buxton, at the
distance of six miles, is the village of Taddington, with an
elevation of 1122 feet. The almost contiguous high grounds
of the village of Fairfield, flank and rise above the town of
Buxton on the north-east; beyond which, at greater and
greater distances, rise the higher and higher grounds of Peak-
Forest, Mam Tor, and Kinderscout. These surrounding
ranges of more elevated ground, not only protect Buxton in a
considerable degree from the more severe effects of prevalent
winds, but the more or less steep ascents and declivities of
the sides of the oblong basin represented by the valley of
Buxton, offer a great variety of scenic beauties. Several
hundreds of acres of the valley, to the west and south-west

of the town, present swells and undulations of great natural capability, much of the land sloping gently towards the south. These grounds have been partially turned to much account, more particularly within the last three years. The Buxton park occupies 120 acres of this part of the valley; and contiguous to the park, the principal public pleasure grounds, gardens, and plantation walks have been made. The higher grounds, which surround the valley on all its sides, are for the most part crowned with plantations, which not only serve to enrich the landscape, but must assist greatly in tempering the severity of the mountain winds.

For a full comprehension of the Buxton district, with its roads, and hills, and valleys, and objects of interest, reference is made to the excellent accompanying map; and likewise to an equally sufficient plan of the Buxton park, walks, and pleasure grounds, which has been prepared in illustration of this work. As both the map and the plan have been necessarily drawn to a scale, the distances between the different places, and the extent of the different walks, may be readily estimated.

Buxton is situated on the south-western edge of an extensive formation of mountain-limestone. The formation presents the usual characteristics of the secondary limestone. The surface of the country is remarkably undulating; broken in the course of the streams into bold ravines, which are bounded by lofty and precipitous crags, having deep, time-worn, perpendicular fissures, with frequent horizontal cracks, often of great length, and as straight as if formed by art. These cracks often extend deeply beyond the mere surface of the rocks; and in many places, time or art has removed in part the upper layer or layers, and left broad shelvings of rock, which illustrate very well this character of the formation. In the instance of a well known rock in this formation, Chee Tor, the appearance is as if the upper part of the vast mass had been carefully cut off the subjacent layers, and accurately replaced in the same position. The long and winding valleys of this formation, with bright trout-streams rippling and tumbling over their rocky bottoms,—with beetling, precipitous, clefted, and time-worn crags, of pale-grey colour, bounding their sides,—and mountain-ashes, yews, pines, hazels, and thorns, partially clothing, without concealing, their romantic and various ruggedness,— while the anemone, orchis, saxifrage, forget-me-not, &c., embellish them with minuter features of beauty,—contrasting, as these valleys do so very remarkably, with the large-featured upland scenery of this district, on which the eye wanders for miles, until in one or two instances it almost distrusts the evidence of its impressions, and on which the lights and shadows of the clouds are often mapped with a curious and exquisite distinctness, and where the distant storm or distant sunshine may be traced at different points in a single view,—cannot but be admitted to give a variety, and character, and degree of scenic beauty and effect to this locality, which can be met with in few places in this country.

One remarkable characteristic of the mountain limestone formation, is well exemplified in that of the Peak of
Derbyshire. It contains many large natural caverns. These
caverns, the more important of which are at Castleton,
Matlock, and Buxton, are entered by natural arches or
fissures, at different elevations of the sides of the hills in
which they are situated, and lead to alternate passages and
chambers, which differ much as to height, windings, and
length; the chambers being in one or two instances of
palatial size, and of noble height and proportions;
in some cases roofed with a flat surface of rock, in
others with arches of different forms and sizes. In the
great Peak cavern, at Castleton, these arches, from their
height, span, proportions, and harmony as to character
and extent with the chambers which they canopy,
fill the mind with a sense of grandeur and beauty,
scarcely inferior to that produced by the interiors of the
larger cathedrals. In some instances, the constant dripping
of water from the roofs of these caverns, charged with
calcareous matter,—in others, the constant oozing and
welling of such water over large faces of the rocky sides of
the caverns, have, in process of time, formed stalactites of
great size and curious variety, or produced surfaces of
crystalline character. In the Blue-John cavern, at Castleton,
the crystalline surface resembles a great cascade, and presents,
when well lighted up, a remarkably intricate and beautiful
variety of surfaces and reflections. It is remarkable, and
adds much to the effect of these caverns, that a stream of

water passes through the larger number of them. Some
geologists have expressed an opinion, that such streams may,
during the lapse of ages, have produced these great excava-
tions: but this is not possible. It would be difficult to
infer such an amount of effect from a flow of water, that is
in general small and unimportant; although the influence of
such an addition, with its darkness and its murmurings, on
the character and impression of the caverns upon those who
explore them, can scarcely be over estimated. And, moreover, there are chasms, and arches, and caverns, in this
formation, which show no evidence of having been water-
channelled at any time; and, therefore, there can be no
doubt, that the whole of these have been equally the effect
of disruptions, probably the immediate consequences of
volcanic action.

That volcanic action has been in extensive operation in
this district, at some remote period of time, is not only shown
in this way; and not only, in having probably formed the
fissures, through which the tepid mineral waters of the district
find their way to the surface; and not only, in the displacements, and shatterings, and extensive disruptions of the
limestone strata; but evidence is given, that molten rocks
have, in some places, overflowed the ordinary strata,—thus
covering, or underlying, or mixing with the limestone
which had not been acted upon by fire. Sir Henry T. de
la Beche, in his great work, "The Geological Observer,"
says—"In Derbyshire the observer will again see igneous
rocks associated with ordinary deposits; in this case with

limestone, known as the carboniferous or mountain limestone, in such a manner that their relative geological
antiquity can be ascertained. Careful investigation shows
that in that area, at least, and probably much beyond it
(beneath a covering of the sands, shales, and coals, known
as the millstone grit and coal measures), and after a certain
amount of these limestones had been accumulated, there had
been an outburst and overflow of molten rock, irregularly
covering over portions of them. And further, that after this
partial overflow, the limestone deposit still proceeded;
probably spreading from other localities, where the conditions
for its accumulation had continued uninterruptedly. Occasionally water action upon the igneous products may be
inferred prior to the deposit of the calcareous beds upon
them, if not also a certain amount of decomposition of the
former, the limestones immediately covering them containing
fragments (some apparently water-worn), and a mingling of
the subjacent rock, such as might be expected if calcareous
matter had been thrown down upon the exposed and
decomposed surfaces of the igneous rock. In some parts of
the district another outflow of the same kind of igneous
rock again took place, and was again covered by limestone
beds, so that in such portions of the area, two irregularly
disposed sheets of once molten rock are included among
the mass of the limestone beds." The same excellent
authority adds, that, of these igneous rocks, locally known
as toadstones, "natural sections (many of which are
excellent) and mining operations show that as regards

thickness these overflows vary considerably, so much so as
to aid the observer in forming some estimate of the localities
whence the molten matter, when ejected, may have been
distributed around." . . . . "In the case of
Derbyshire, though there may have been a removal of a
portion of the igneous beds by the action of water upon
their exposed surfaces (and an attentive examination of the
upper overflow shows a quiet adjustment of the limestone
beds formed upon it), no deposits resembling the ash and
lapilli beds above-mentioned as found in Devon and Cornwall, Wales and Ireland, have yet been detected. There is
no evidence showing an accumulation of ash and cinders in
the manner of subaërial volcanoes. It may readily have
happened, therefore, that the igneous matter was thrown out
in a molten state, without any accompaniment of ash and
cinders; and this might have taken place as well beneath
the level of the sea as above it." These are some of the
wonderful phenomena of primeval nature; and they furnish
an interesting illustration of the simple way, in which they
may often be studied and explained. They show the
gradual and perhaps slow formation of the limestone rocks
at the bottom of the sea, and the occasional disturbances
produced by volcanic outbreaks, modified in their degree
and effects by the superincumbent ocean, which would
probably not only moderate the violence of such action, but
circumscribe its effects; the deposition and accumulation of
the calcareous strata being only interrupted during the time
that the volcanic outbreak might be going on, and possibly

to no very great distance beyond the immediate locality of such outbreak. "Upon examining the structure of the igneous rock, it is found to be partly solid, and confusedly well crystallised, a compound of felspar and hornblende, with, sometimes, sulphuret of iron. It is partly vesicular, in some localities highly so; the vesicles, as usual, filled with mineral matter of various kinds (carbonate of lime, as might be expected, being very commonly present), where the rock has remained unaffected by atmospheric influences, but exhibiting the original and vesicular state of the molten rock where these have removed the foreign substances in them. In some localities the scoriaceous character of the rock is as striking as amid many volcanic regions of the present day. Like more modern igneous products, also, it will often be found decomposed in a spheroidal form. There is an example of this decomposition at Diamond hill, on the south side of Millar's dale, where the concretionary structure has been developed somewhat on the minor scale, and the size of the spheroidal bodies is about that of bomb-shells and cannon-balls."—*Sir Henry T. de la Beche.*

The outflow of these igneous products in the district more immediately around Buxton, may be compared to the tortuous meanderings of a mountain stream. These meanderings of toadstone extend from Fairfield to the Water-Swallows, where there is a much broader and more considerable outflow; the narrower meanderings of the toadstone continuing thence to Peak-Forest, and thence to Tideswell, Wormhill, Millar's dale, Litton, Ashford, Chelmorton, and

Buxton. The toadstone varies much in its density and general physical character; but it always presents the distinctive difference from the limestone, which likewise varies much in its density, that the action of fire has deprived it more or less entirely of the stratified character of rocks formed by deposition. In different places and specimens, the toadstone shows varying evidence of igneous action, from a friable, light, and porous, lava-like tufa, to a dense, and much more fully vitrified, and compact rock.

The mountain-limestone contains a great variety of fossil shells; and such may be said to constitute a large proportion of the rock and marble of which it is composed. The common grey marble of this district, is evidently altogether composed of dense masses of shells; and a dark-coloured marble, known as the bird's-eye marble, is in a great degree composed of shells. It needs no taste for geological pursuits, and but little acquaintance with the wonders, as to the formation and early history of our globe, which geology teaches, to make this a matter of curious interest to every one. The limestone rocks, in all directions in the neighbourhood, show, on their abrupt and craggy surfaces, dense masses of these primeval shells; indicating a time when this high range of country was submerged in ocean; and when, as it should seem, by the agency of myriads of these marine creatures, such masses of rock were altogether or in large degree produced. These fossil shells differ essentially from those of the existing species; and differ from one another as much in size and form, as the marine shells of the existing species differ from one another. As to size, some of the fossil shells are several inches in diameter, and others are so small as to be altogether invisible to the naked eye. Lamarck well says "in producing living bodies, what nature seems to lose in size she fully regains in the number of individuals, which she multiplies to infinity, and with a readiness almost miraculous. The bodies of these minute animals exert more influence on the condition of the masses composing the earth's surface, than those of the largest animals, such as elephants, hippopotami, whales, &c., which, although constituting much larger individual masses, are infinitely less multiplied in nature." As the coral reef, rising in the midst of the ocean, in our times, comes at length to emerge above the level of the waters, and to form a new land, on which birds may alight, and alluvial soil be formed, and to which seeds may be wafted, and where vegetables may grow and flourish; and all this marvellous sequence, involving the formation and completion of a new and habitable country, be referable to the labours of myriads of coral insects; so, by means of myriads of marine creatures, requiring and producing these coverings of shells, was this formation of secondary limestone in great degree produced,—to be at length upheaved, probably by volcanic influence, from the bed of the ocean,—to be partially vitrified by the heat, its organic structure being so far destroyed, and a crystalline or an amorphous character substituted for it,—to become partially mixed with products of volcanic action,—in part to form rugged and broken masses of precipitous rock, to be worn by the storms of ages,—in part to show marks of disrupted stratifications, the shakes and displacements, which tell even now, in the strongest language, of the convulsions by which such masses were upheaved,—in part to become extensive surfaces of undulating country,—in part to form the rocky sides of valleys, between which the streams from the mountains may find their way to the sea. Such are the rocks, the uplands, and the valleys of the Derbyshire limestone.

The sojourner who can gather such food for thought in his walks about the neighbourhood of Buxton, has before him in this locality abundant additional materials for his enquiries. Leaving the mountain-limestone formation, on the very edge of which he finds himself when he passes to the north and north-west of the town,—and crossing the narrow bed of shale, which he does on commencing the ascent of the Manchester road,—he immediately steps to the adjoining formation of millstone-grit, which tells of a less remote period in the world's history. In a quarry of valuable stone for building purposes, about half a mile from the town, on the right hand side of the road, are occasionally found the fossil remains of fruits and monocotyledonous stems, which show that, at some remote period, the climate of these now colder regions of the world, must have been at least as warm as that of the intertropical countries of modern times. These fruits and stems show, that plants which only grow and flourish within the limits of the torrid zone, must at one time have attained a large size in this locality. How strange, and yet with what a strong probability of truth, to think that possibly this gritstone formation was, at some remote period, part of a land of much lower level than that which it now occupies, the temperature of which was that at which palms and the like can grow and flourish; while the adjoining formation of secondary limestone was at the bottom of the sea, or perhaps in process of being formed by myriads of shell-fish!

The contiguity of the limestone and gritstone formations affords much matter of curious observation, as to the difference of vegetative power and character of these different soils.

The moorland character of the uncultured higher grounds of the gritstone formation,—the peat soil,—the existence in many places of such considerable thicknesses of bog-earth overlaying the gritstone, as to have contained large trunks of trees completely buried and preserved, for periods probably beyond recorded time,—illustrate remarkably the very different early history of the limestone and gritstone formation. At those remote periods, these parts of the gritstone formation must have been covered with a dense vegetation; layer upon layer of which, buried by new growths, to be in turn buried by successive growths, at length formed great depths of impervious bog-earth, which retain the rains in a chill and unproductive excess of moisture, and form a surface only capable of supporting heaths and kindred plants, until subjected to such dressing and drainage as alter its character and condition. At these remote periods of

time, the limestone may have been at the bottom of the sea,
or in the last stages of its formation; and at all events must
have been so far differently circumstanced, that it had no
vegetative growth of similar character to that of the
gritstone formation.

The vegetation of the pasture lands differs much on
these formations; there are marked differences in the
broader features of the landscape; and some trees, and
plants, and wild flowers, which thrive on the one, do not
thrive on the other. These differences in the characters
and productions of the two formations, are especially remarkable in their respective valleys. There is a valley on the
gritstone formation, which begins at a short distance from
Axe Edge, and extends several miles. This valley, called
the vale of Goyt, from the mountain stream—the Goyt—
which runs through it, exhibits throughout its course a
remarkable richness and variety in its vegetative growths.
This is in part due to the gritstone detritus, which constitutes
necessarily much of its soil, and in part to coverings, or
admixtures, or detritus, of peat or bog-earth, of varying
thickness and proportion. Trees grow with great rapidity
on the sides of this valley. There is scarcely a wild fruit
which grows in any part of these kingdoms, that is not to be
found growing in this valley, or on the adjoining uplands and
moors,—from the cloudberry, clusterberry, cranberry, and
bilberry of the moorland, to the blackberry, strawberry, and
raspberry, of the valley and its sides. But, to return: there
is a great and readily observable difference in the character

and general appearance and form of the surface, in the shape
of the hills, in the appearance of their sides, curves, and
eminences, and in the whole character of the vegetation of
the gritstone and limestone formations.

There is a magnificent and much broader valley, within
the same distance from Buxton, on the north; being divided
from the valley of Buxton by Comb's Moss. The town of
Chapel-en-le-Frith, which is six miles from Buxton, is
situated in a part of this wide and undulating valley, or
extensive basin, which consists almost entirely of the gritstone formation. There are few finer scenes, than the view
of this valley from the north-western edge of Comb's Moss,
at the distance of somewhat less than three miles from
Buxton. The explorer may turn off the Manchester road to
the right, at the first milestone from the town, follow the
bridle-road for about half a mile, and then ascend the higher
grounds on the right. To a stout pedestrian, however, the
whole of this valley, as well as that of the Goyt, is well worthy
of being explored.

At a distance of about half a mile from Buxton, on the
north-east, is the hamlet of Fairfield, with its fine commanding upland position, its church, and its extensive common
—the old Buxton racecourse. The road from Buxton to
Fairfield is a steep ascent; presenting on the left, a very
good view of the whole valley of Buxton, backed and begirt
by Axe Edge, Grin Edge, and Comb's Edge; with Lower
Buxton, and its Crescent, and church, and the adjacent park,
occupying the centre of the scene. The village of Fairfield

is prettily situated on this upland; and beyond it lies the common, which affords admirable ground for horse-exercise. The road which leads to Chapel-en-le-Frith, passes at right angles, almost immediately beyond the common, part of the old Roman road called Batham gate. The undoubted antiquity of this road, together with the name it has immemorially borne, serve to support the ancient use and importance of the Buxton baths.

If this old road, with its less evidently ancient continuations, be followed for about two miles, the so-called Marvel-stones will be seen on the right. This is a curious and somewhat extensive cropping out of limestone rocks, which are raised two or three feet from the surface. The less zealous explorer will however hardly think himself repaid by their appearance, for the trouble of his journey to the spot.

Immediately beyond the Marvel-stones, lies the small mountain village of Peak-Forest, with its chapel, which is said to have enjoyed, so recently as in the course of the last century, the celebrity and supposed privileges of an English Gretna-Green. Very near to Peak-Forest village, there is an extraordinary natural opening or fissure in the limestone, called Eildon-Hole. The depth of this fissure, and its irregularity, must be great; inasmuch as, on throwing stones into it, they often fall and rebound from side to side, until the reverberation comes to be heard more and more faintly, the sound seeming to be at length lost in the greater and greater distance. There may be some degree of deception in this matter, owing to the echoing effect of the reverberation in the contracted and rocky channel; but it seems to be probable that the depth of the chasm is really very considerable.

About a mile beyond the junction of Batham Gate with the Chapel-en-le-Frith road, at Dove-Holes, is one of the remarkable water-swallows, of which several are met with in this district. A larger or smaller stream of water descends by a fissure into an under-ground natural channel, and emerges from the surface at a greater or less distance,—in some cases said to amount to a mile, or even more.

One mile beyond Dove-Holes, this road joins the main road, which leads from Chapel-en-le-Frith to Castleton. The main road descends rapidly towards Chapel-en-le-Frith, which is about a mile and a half from the junction of these roads. The small town of Chapel-en-le-Frith is prettily situated and sheltered. Immediately beyond the town, the valley in which it is situated opens out to a considerable width, presenting bold and fine elevations towards the north and south, and enclosing beautiful and productive lands on both sides of the road. This road joins the high road from Buxton to Manchester, about three miles from Chapel-en-le-Frith, and six miles from Buxton, at Horridge end, and close to the hamlet of Whaley.

To the north of Chapel-en-le-Frith, are the districts and towns of Hayfield and Glossop; and to the east of Hayfield, is the great range of elevated country, which is dignified more especially by the name of *the Peak*; having Kinder-

scout on its western, and Ashop Moor and Edale on its
eastern extremity,—the higher grounds having an elevation
of nearly 2000 feet above the level of the sea. The
beautiful valley of Edale, than which even this district has
few finer scenes to offer, separates this extensive range of
high lands from Mam Tor, which, although only 1709 feet
above the sea-level, from overlooking Edale on the north, and
the more extensive valley of Hope on the south-east, is often
considered to be, as would be implied from its name, the
greatest of these eminences. Immediately at the foot of
Mam Tor, lies the old village of Castleton, crowned on its
southern side by the smaller, but steep and commanding
eminence, on which are the ruins of the castle of the lords
of the Peak in the olden times. The view from these ruins
is extensive, and very fine and varied; and indeed the whole
district supplies such a number and variety of scenes, that
every half mile of a journey furnishes a new and extensive
picture.

Close to the village of Castleton, is the great Peak cavern,—
the most remarkable of all the Derbyshire caverns,—which is
entered by a natural arch, forty-two feet high, and one
hundred and twenty feet wide; this imposing hall of entrance
being three hundred feet in depth. Beyond this hall, a
narrow low passage, almost separated from the further
interior by water, which is either crossed by an artificial
foot-path or by means of a boat, conducts the explorer into
a spacious cavernous chamber, some parts of which are
estimated to be two hundred and ten feet in width, and

one hundred and twenty feet in height; the whole being enarched, with a magnificence of general effect, and a beauty and variety of detail, which baffle all description.

A lead-mine, no longer worked, called the Speedwell mine, is another of the wonders usually explored by the visitor to Castleton.

The Blue-John mine, whence the curiously beautiful spar called Blue-John is obtained, is well worthy of a visit. Vast spaces of the sides of this cavern are covered with sparry incrustations of great variety, reflecting most beautifully the lights of the candles and crimson and blue fires, by which the cavern is illuminated by the guides.

The whole of the valley to Hope and Hathersage, and the great extent of hills and moorlands to the north, east, and south, are well worthy of being explored, and cannot fail to excite and elevate any and every mind, and fulfil abundantly any amount of expectation.

The traveller, in going from Buxton to Castleton and back, will act wisely to go on the one occasion by the road which passes the foot of Mam Tor, and on the other to pass through the Winnetts or Wind-gates. The view through these great rocky portals presents a dioramic scene of magnificent extent and beauty.

About five miles from Buxton, by the side of the road to Castleton, at the upper part of the valley of Bar-moor Clough, through which the road passes, is the most remarkable of the intermitting springs of this district. It is called the Ebbing and Flowing Well. The frequency with which

this intermittent flow occurs, depends upon the amount of
rain which may have fallen recently. After much rain, the
flow may be as frequent as every ten or fifteen minutes.
The quantity of water poured out at a time must be considerable. The ebb and flow may be due to a curved conduit,
through which the supply of water has to pass. One limb
of such conduit might become gradually filled with water
as it drains from the surface; at the same time the
water would rise to the same level in the other limb
of this natural syphon; and when the second limb had
become filled to its further extremity, the flow would
take place, and continue until both limbs of the conduit
were emptied, when the flow would cease, and the curved
conduit have to be again filled.

The whole course of the Derbyshire river Wye, from
Buxton to its junction with the Derwent, at the village of
Rowsley, beyond the town of Bakewell, presents a great
variety of valley scenery of remarkable beauty. The road
from Buxton to Bakewell passes through Ashwood dale,—
the nearest of these valleys to Buxton. This valley is
rather more than four miles in length; and the road passes
close to the right bank of the river about three-fourths of
this distance. Near to Buxton, the valley is bounded by
abrupt limestone rocks of considerable height, and much
bold and rugged character. Several smaller valleys
open from Ashwood dale; and one of these, from its
remarkable and picturesque beauty, deserves to be particularly mentioned. This is Sherbrook dell, opposite to the

first milestone from Buxton. The sides of this dell are extremely abrupt and lofty rocks, which hem in the narrow gorge completely; and as the ravine bends suddenly within a few yards from the road, the explorer finds himself at once surrounded by much untouched and majestic natural beauty: the rapid and bubbling streamlet, by which its bottom is channelled in the winter time, and after heavy rains,—the little cascade which tumbles into the dell at its upper end,—and the wild plants and shrubs by which every cranny and crevice are taken possession of,—all serve to embellish this little dell very much.

The greater part of Ashwood dale is planted on both sides, almost to the summits of the rocks. From beyond Blackwell mill, four miles from Buxton, there is a short extent, with scenery of more open and wilder character, as far as to a sudden turn of the valley and the river, with steep and rocky buttresses, marking the entrance to Chee dale. The magnificent perpendicular mass of almost circular crag, called Chee Tor, is in Chee dale, on the right bank of the river. This vast mass of rock is of considerable height, but necessarily seems to be of greater height than it is, from its perpendicular sides, which are as straight as if cleft with care by the hand of man. The curious horizontal fissure near the summit of this rock has been already noticed. The perpendicular cliff towers above the dale on one side; the bright stream occupies the bottom of the valley; on the left, the hill side is embellished with scattered and overhanging trees and bushes; and an appearance of isolation is given to the scene, by a bending of the valley to the left and then to the right, in order to skirt the rounded projection of the Tor,— the valley being thus shut in on all its sides.

Passing from Chee dale and its great tor, there is a steep but practicable foot-path up to the village of Wormhill; and from the upper part of this path, a fine view across the valley is obtained. The river now passes below Priest-cliff, a gently sloping and rounded hill, which is for the most part planted, as are the sides of the further valley, which here takes the name of Miln-house or Millar's dale. This is a much more open valley, with sloping sides; patches of plantation and juttings of limestone rock varying the surface. The river is here of a considerably wider and more imposing character; and the scenery is less like that which commonly characterises the limestone valleys, and is more like the valley scenery of other parts of England. At the end of about two miles, however, the valley again contracts; the river becomes again confined within narrower bounds; the sides of the valley, although clothed with trees, are again more precipitous, and the characteristics of the limestone formation are again strongly exemplified. The course of the river is here unusually tortuous; and, emerging from this narrowed valley, it enters the broader and more slopingly-sided valley of Monsal-dale, with natural plantings of hazel, &c., and a great degree of richness and beauty. After a course through this valley of two or three miles, the river again meets the high road opposite to the eighth milestone from Buxton to Bakewell.

At the third milestone from Buxton, the road to Bakewell, unfortunately made to quit the level of the river Wye, ascends rapidly to the high grounds of these elevated lines of country. Some little compensation, however, is given for the scenery left behind, and for all that which is commonly thus left unseen, by a wide and varied range of scenery on the left; the districts of Blackwell, Wormhill, and Tideswell, being overlooked from the road; and, on nearing Taddington, which is six miles from Buxton, the higher grounds of Chelmorton Low are on the immediate right; and the road is so much higher than the village of Taddington, that a view is commanded of the high grounds of East Moor, at a distance of ten or twelve miles. From the village, the road rapidly descends, and enters the valley of Taddington, which is bounded on both sides by lofty elevations of much beauty, and some occasional grandeur. The sides of the dale are clothed by natural plantations of hazel, hawthorn, &c. After a descent of two miles, the road again joins the course of the river; and passing the end of Monsal dale, and crossing and re-crossing the stream, it leads, through the pretty village of Ashford, to the neat and clean and pleasant town of Bakewell.

Beyond Bakewell, the road still maintains its position by the banks of the Wye, through the vale of Haddon; passing the fine old mansion, Haddon Hall, on the left, about two miles from Bakewell. Haddon Hall deserves a volume of description; it is almost unique, as an untouched specimen of the homes of England's aristocracy, in the olden times.

Situated on a fine terrace, on the side of the broad valley—with its bridge, and old gateway, and court-yard, and many windows, and irregular walls, and numberless rooms (for the most part small and very irregular), and old chapel, and buttery, and kitchens, and hall, and long gallery, and magnificent views from narrow windows, and lofty parapets, and garden terraces,—Haddon Hall deserves all the attention which it receives from painters and tourists.

Beyond Haddon, the road soon leads to the cheerful village of Rowsley; and the Wye at this point loses its identity, and becomes involved in the larger stream of the Derwent.

Crossing the river, and proceeding northward, towards the village of Edensor, Chatsworth park is soon reached —the princely domain of the Duke of Devonshire. Chatsworth House is remarkable for its size,—its adaptation to the scenery which surrounds it, — its back-ground of dark woods, which shelter an arboretum of botanical and numerical value and importance,—its gigantic and costly fountains and water-works, — its great art-created rock-works, — its large conservatories and orchid-houses, —its extensive, and varied, and art and taste and science-serving gardens and pleasure-grounds,—its imposing Italian elevations,—its princely suites of rooms,—its most choicely-filled sculpture gallery, — its paintings and drawings by great masters, ancient and modern,— its august library, collected by successive generations of noble lovers and patrons of letters, and in part by the great philosopher,

who gave additional distinction, even to
Cavendish; these great treasures, and this
house, and its belongings, being most kin(
be seen by the public at large.

From Rowsley to Matlock, the road foll
course of the river Derwent, through the be
valley of Darley dale, in which breadth, ar
variety, fill and satisfy the eye and the min
way between Rowsley and Matlock, on t
road, is Darley dale church, by the side
one of the oldest yew trees in England,
growth of many centuries.

Matlock village is eight miles from
between the village and Matlock bath, wh
usurp the name of Matlock, there is a dista:
miles. The road passes near to the river, wit!
and deeper stream; and the valley rapidly b
by loftier and more precipitous rocks, until
with its accumulated picturesqueness of outl
its massive rocks,—its romantically placed
all elevations, and in positions which seem t
when they are first seen; with its walks, a
and caverns; with its well appointed hot
houses; with its tepid natural water ar
degrees,—form a watering-place of much at

Having passed Matlock bath, the road e
extraordinary portion of its course, through
portals, to the almost contiguous village o!

thence, still occupying one or other bank of the river, passes
to the town of Belper, an important seat of the cotton
manufacture; goes through the pleasant village of Duffield,
and reaches Derby, with its fertile surroundings, at the
distance of 38 miles from Buxton.

Beyond upper Buxton, on the south side of the town, is
the old road to London, viâ Ashbourne and Derby. This is
the most exposed and least interesting of the roads near
Buxton. Passing the beautiful rising grounds of Stadon,
near to the town, and leaving the village of Chelmorton on
the left, about four miles from Buxton—Chelmorton being
situated at the foot of Chelmorton Low, in an open valley—
the scenery of the adjoining county of Stafford being shut
out by the rising grounds on the right—the road passes
over uplands, of bare and tame character, to Newhaven; and
thence, with little improvement in the scenery, until some
fourteen miles from Buxton, when the village of Tissington
on the left, with its old trees, and enriched Old-English
character of scenery, and the road to Dove dale on the right,
would atone for a much less inviting intervening tract of
country.

If, instead of following the high road just noticed,—the
merits of which, for excellence of condition, deserve a passing
word of praise, even in a district where all the principal
roads are maintained in the best possible state,—the traveller
diverges from it to the right, when between two or three
miles from Buxton, he ascends at once the invidious range
of higher ground, which separates him from the scenery of

Staffordshire. He passes, on his right and left, many oddly shaped, and bold, and picturesque hills. He may go through the village of Earl Sterndale, ascend the conical hill called High-Wheeldon, and see all these hills before and below him, and an extensive and picturesque reach of the county of Stafford, lying on the other side of the river Dove, which divides Derbyshire from Staffordshire. The more distant of the Staffordshire scenery is divided from that which is nearer, by successive ridges of hills, over which the eye travels, and upon which the lights and shades of the clouds produce the most picturesque and rapid changes. A descent and ascent, through interesting and ever varying valley scenery, with a crossing and re-crossing of the river Dove, here a stream of no pretension, leads to the little market town of Longnor, at a distance of six miles from Buxton. Passing thence to the left, through a picturesque valley, which winds around the base of High Wheeldon, the traveller regains a road which leads to the village of Hartington. Hartington, by the more direct road, is about nine miles from Buxton. There is little scenery worthy of remark on the road from Earl Sterndale to Hartington, until within half a mile from Hartington, whence a higher level of road commands a view of the river Dove, with a much improved description of scenery bordering its course. Hartington is an unpretending and quiet village; and, within about half a mile, the river Dove passes through the first part of the remarkable scenery, which has rendered it so famous. Beresford dale is a little gem of beautiful scenery. The more matured beauties of

D 2

that part of the river's course which is more strictly called
Dove dale, if here somewhat less bold in character, are
crowded together into a small space, giving the sense of
finished beauty which an exquisite miniature conveys, and
which may compensate in some degree for any deficiency
in the boldness and character, which might distinguish a
painting of more pretension, and on a larger scale.   Here,
too, is the fishing-lodge, which was erected for the accommodation of the venerable Izaak Walton; here is the
domain that belonged to his disciple, and friend, and
expounder, Charles Cotton; here is the stream, from which
those lessons in angling were obtained, and by the banks
of which those thoughts and maxims and gossip were
formed, and embodied in simple and quaint phrase, which
still serve to edify and please all the lovers of nature, as well
as those who practise the "gentle art"; and every one may
well feel that this is indeed a fitting theatre for such
thoughts and musings.

_ Some few miles of bare scenery border the course of the
Dove from thence to Dove dale.   Dove dale is separated
into several almost distinct portions, every one of which is
distinguished by its own peculiar and characteristic beauties.
The first of these is a somewhat open valley, with a rippling
and shallow stream, and grassy banks and bottom, and
shelving and less bold sides,—with but little of rocky and
limestone character, until the eye reaches to the higher
portions of the mountain walls. This is the less adorned
hall to the more enriched scenes beyond. Passing over

some higher ground, which serves to shut out this first
compartment from that immediately beyond, the eye is
arrested by a mass of rock, which rises abruptly, standing in
relief, and with much grandeur, on the right side of the
valley. On the left, a little beyond this grand mass of
limestone rock, is an expanded arch, of fine form and
proportion, leading to a shallow cavern. Beyond this, on
the left side of the valley, is a mighty and marvellous specimen
of the peculiarities and capabilities of the mountain limestone. A mass of rock, standing out boldly from the
mountain side, at an estimated elevation of between two
and three hundred feet from the bottom of the valley, is
completely perforated by an arch of some yards in depth,
and said to be about forty feet in height, and eighteen
feet wide. Through this archway is a space, open to the
sky, which might be likened to the small court-yard of a
mountain stronghold; and which leads to a narrow cavern
in the higher hill-side. This curious archway, which has
become detached from the further cavern, situated as it is
at so considerable a height, admitting the light of day freely
through it, and presenting the view of the space and cavern
beyond it, is one of the most picturesque of the rocky
wonders of the limestone formation. The view of the valley
from above, looking through the archway from the upper
cavern, is sufficiently beautiful to repay fully the toilsome
ascent by which it has to be attained.

The dale immediately beyond becomes much narrower;
the sides become precipitous and rocky; the river becomes

narrowed, and less quiet in its character; and enters a narrower and darker gorge between two great rocky portals. On one side is a column of insulated rock, which rises abruptly, and in massive grandeur; on the other side is a bold mass, projecting from the side of the valley. What a scene of "hurly-burly," and what gigantic action, must have produced and attended the dislocation and upheaving of these mighty masses; and what a tale this scene tells, in the midst of the beauty which is now so solemn and so still, of the agency by which the earth's strata were made to produce the diversified surfaces, so needful for the wants, and conducive to the uses, the health, and the happiness of the human race!

The valley below has again a more open and more enriched character; with a more quiet and broader stream, bounded by more sloping hill-sides; broken at intervals by masses of rock, scattered in vast fragments, or projecting, as though they had only just escaped from being hurled into the valley which they overhang.

The town of Leek, a principal seat of the silk-manufacture, is at the distance of twelve miles from Buxton, on the south-west. The road rises rapidly from the valley of Buxton, passing over the ridge of elevated ground, of which Axe Edge is the highest point. The first part of the road is wild and bold in its scenery; the pointed and oddly-shaped hills, near Longnor, lying at some distance from the road, on the left; the right being bounded by the higher ground of the ridge of Axe Edge. When the summit of the high ground is at length attained, an extensive view is presented; and from thence to the town of Leek, the scenery is of commanding and varied character.

Branching from the Leek road, on the right, rather less than two miles from Buxton, is the road to Congleton, a small town, at the distance of fifteen miles. The road is wild, and less interesting than the road to Leek, or than that to Macclesfield.

. Macclesfield is at the distance of twelve miles from Buxton, on the west. The highest part of the road, close to a small road-side inn (the "Cat and Fiddle"), offers a view which is circumscribed in breadth, but which extends in length to a distance of forty or fifty miles. The river Mersey, near Liverpool, may be seen from this point, when the air is free from haziness, as after rain; the looking at objects so distant being even painfully fatiguing to the eye. From this point to Macclesfield, the road descends; offering an extensive view over this part of Cheshire, and leading to an idea of the town of Macclesfield which must be admitted to be beyond its deservings. Macclesfield is a well-known and very important seat of the silk-manufacture.

On the north-west of Buxton is the city of Manchester, at the distance of twenty-four miles. The road from Buxton ascends for a distance of two miles, having the valley of Chapel-en-le-Frith on the right hand, and that of the river Goyt on the left; no part of the one, however, and but little of the other, being visible from the road. The village of Taxal, and, almost contiguous to it, that of

Whaley, are six miles, and that of Disley is eleven miles from Buxton. Close to the left of the village of Disley is the extensive park of the Lyme Hall estate, the property of Mr. Legh's family during several centuries. The Hall, to which the public are kindly permitted to have access, forms one of the great attractions of the district. The house is interesting from its associations, and contains much that is of historical, and much that is of intrinsic interest.

Six miles beyond Disley, or seventeen miles from Buxton, is the town of Stockport,—which, only seven miles from Manchester strictly so called, is becoming little else than an extension of that great metropolis of the manufacturing districts of Lancashire and Cheshire.

As will have been inferred from the account which has been given of the Buxton district, the walks and drives in the more immediate neighbourhood of the town are interesting.

The walks opposite the Crescent, already mentioned as having been formed on the side of an originally unsightly cliff by Sir Jeffery Wyatville, offer a valuable resource to those who are more especially invalided, from their proximity to the principal hotels and lodging-houses. These walks are cut out of the limestone rock, and are accordingly remarkably dry. Arranged in a succession of terraces, a series of level walks is obtained, with the advantage of giving occasion for climbing at pleasure, or as lameness may diminish, or strength increase, to a higher and higher terrace walk; and thus these successive terraces have

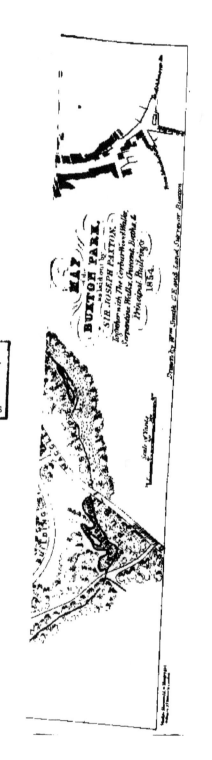

MAP
of the
BUXTON PARK,
as laid out by
SIR JOSEPH PAXTON
Together with The Corbar New Walks,
Serpentine Walks, Crescent, Baths &
Principal Buildings
1854.

Scale of Yards

Drawn by W.m Smith C.E and Land Surveyor Buxton

long been popularly recognised, as supplying indications of
restored power and capability, in regard to the crippled
limbs and enfeebled state of those resorting to Buxton for
the use of its waters (see the plan of the walks, park, &c.).

Almost contiguous to these walks, at the west end of the
Crescent, opposite to the Hall and the Square, are still
more extensive walks and pleasure grounds, maintained and
kept in order for the free use of the public. These walks,
which have been much extended and improved of late, are
carried through a long belt of plantation, on both sides of
the river Wye; the stream being crossed and recrossed by
rustic bridges, diversified by water-falls, and in other ways
subjected to the requirements of ornamental grounds.
Thoroughly drained throughout the whole extent of these
plantations, the walks having been rendered remarkably dry
by this means, and by having been carefully made; and
the most having been made of an originally great capability;
these walks are sheltered and pleasant, and are much resorted
to. These plantation walks, which are sometimes called the
Serpentine-walks, and sometimes the Winding-walks, furnish
a circuit of dry, well-gravelled, and well-kept foot-paths,
of considerably more than a mile in extent.

Only separated from these walks by the high-road to
Macclesfield, is the Park, which occupies more than a
hundred and twenty acres of greensward, sloping towards
the south, and with walks and drives carried through it for
the use of the public. Close to the town, on a rising and
well-drained upland, well exposed to sunshine and air, near

the lower part of this broad valley, the wide and excellent road around the circle of twenty acres in the centre of the park, is a valuable addition to the resources of the place.

Again : only separated from the park by the high road to Manchester, there is a great extent of walks, of extreme beauty and variety, through a plantation which occupies the site of old gritstone quarries, and covers a great part of Corbar hill-side. Occupying the south side of a commanding eminence, winding through plantations of adequate growth, and traversing the picturesque inequalities of the old quarries, all their rude handiwork covered over long ago with wood, and undergrowth, and ferns, and foxglove, and more recently with rhododendrons and the like, with vistas of Buxton and its valley, and surrounding hills, these walks are the most recent and most picturesque addition to the attractive features of the locality. The terrace-walks opposite the Crescent, the winding walks and pleasure grounds, the roads through the park, and these walks through the Corbar woods, may be moderately computed to supply an extent that must amount to several miles. The more energetic pedestrians should ascend beyond the highest limits of the Corbar wood walks, pass through an upper plantation, and reach the summit of the Corbar hill, which commands an extensive view of Buxton and Fairfield.

The road to Bakewell, winding, as it does, through Ashwood dale, and near to the south side of the river Wye, and leading to much that is interesting, affords a very favourite walk. The road is continued near to the south bank of the stream, for a distance of three miles; but beyond this, there is a practicable foot-path, or rather bridle-road, beyond Blackwell mill; and indeed to the point where the western extremity of Chee dale is shut in by rocks, which abut close upon the edge of the river. A wooden foot-bridge, and but little additional cost, would render the further and very beautiful course of the river accessible from this point, at a distance of only four miles from Buxton.

On the south side of the commencement of the Bakewell road, close to the eastern extremity of Lower Buxton, a stile and foot-path lead through the plantation which covers the southern side of Ashwood dale at this point. The path is carried back again to the road, at the distance of some-what less than half a mile. The road may be left at this point by a narrow foot-path on the south; and this leads, by a continued foot-path, through fields, to Upper Buxton. If the Bakewell road is followed a few yards further than the foot-path now indicated, a road, that is somewhat narrower than the high road, leads to Upper Buxton by Sherbrook and Coteheath, and gives a circuit of rather more than two miles. This, which is commonly called the Duke's drive, is found to be a favourite walk, and likewise a favorite short drive, in much request by those who make use of donkey-carriages and Bath-chairs.

A little distance nearer to Buxton than the first milestone on the Bakewell road, a stile and footpath lead to a wooden bridge across the river, whence a pathway or sheep-track leads up the opposite side of the valley, by the northern end

of a plantation.   On reaching the north-eastern corner of
the belt of plantation, the sheep-track may be left, and the
eastern edge of the plantation may be followed for about a
quarter of a mile or less, when the top of the lofty and
abrupt rocks which bound the northern side of Ashwood
dale will be found to have been reached, and a bird's eye
view obtained of the road, and the river, and all their very
picturesque and beautiful surroundings.   If the sheep-track
shall have been followed, or be now returned to, it will be
found to lead to a green and broad way on the west, called
Tongue-lane, which leads pleasantly over the uplands to
Fairfield, whence Buxton may be returned to by the high
road or otherwise.

   Sherbrook dell, or Lover's leap, opposite the first milestone on the Bakewell road, has been already mentioned,
and should of course be explored.

   By proceeding along the Bakewell road to the bridge
beyond the road-side inn called the Devonshire Arms,
crossing the bridge, and returning through Fairfield to
Buxton by a valley to the left, called Cunning dale, a
pleasant walk of about five miles circuit is obtained; or,
having crossed the bridge referred to, a stout pedestrian
may climb the upland road before him, called Ashe's bank,
cross a field at the top, and reach an old bridle-road, along
which he may return through Fairfield to Buxton, over the
high and open country called Bailey Flat.

   If the Bakewell road be followed through the first toll-bar,
and the steeply-inclined valley immediately on the right be

followed along its bridle-road, a high range of country is reached by Rock Head, or Cowdale; and a foot-path from thence across the fields will be readily found, leading over Stadon, and by Sherbrook and Coteheath, to Buxton, after a circuit of about five miles.

If the Bakewell road be followed a few yards further, and the lodge-gate on the right be passed through, and the road followed to the opposite uplands, the old road to Kingstern-dale is soon reached, close to a small church which has been recently erected; and thence, by turning to the right, after a walk of about half a mile, where the road is crossed by the road from Cowdale to the Ashbourne high road, the return to Buxton may be by turning to the left and gaining the Ashbourne road, or to the right and regaining the Bakewell road, passing by Rock Head, or by walking across the fields over Stadon to Sherbrook.

There are an upper and lower road from Upper Buxton to the first mile on the road to Macclesfield, the one passing by Poole's Hole and Burbage, the other by Wye Head; and returning thence to Lower Buxton by the Macclesfield road, gives a circuit of about two miles. This distance may be shortened by following a foot-path across the fields,—leading, in the instance of the upper road, from Poole's Hole to the plantation walks opposite to the Square and the Hall,—and as to the lower road, from Wye Head to the same place.

There is a pleasant ramble by the foot-path now referred to, from Lower Buxton to the Grin plantations above Poole's Hole, and through these plantations by a cart-road, to the summit of Grin-Low, marked by a mass of loose stones put together to resemble at a distance some ancient ruin. There is a good view of the valley of Buxton from this point.

The second milestone on the Leek road, immediately beyond the toll-bar, is close to the base of the somewhat steep eminence of high ground, called Axe-edge from its lofty and commanding position. From the summit of Axe-edge a good view is afforded of the Buxton valley and its surrounding elevations.

The road to Fairfield, its upland position and extensive common, its fine and bracing air, and the view of Buxton and its valley obtained from it, make it one of the pleasant short rambles near to the town; and there is a foot-path across the fields on the left by which the return to Buxton may be diversified, as well as the routes on the right already spoken of, by which the Bakewell road may be reached opposite to the first or the second milestone, as a shorter or longer walk may be wished for.

There is a long walk of about seven miles, which offers a great variety and beauty of scenery,—from Goyt's Clough, about two miles from Buxton, on the old Macclesfield road, by the banks of the river Goyt, along the moorland bridle-road, into the valley of the Goyt, and as far as Goyt's bridge, —and thence across the bridge, and up the steep old road called Goyt's lane, to the Manchester road, about two miles from Buxton.

Another long walk of much interest, is obtained by leaving the Manchester road at the first milestone, traversing the neglected bridle-road as far as White Hall, descending thence by an old road to the bottom of the valley on the north; thence bearing to the right, and reaching Dove Holes, and thence Fairfield and Buxton, after a journey of about eight miles.

Poole's hole, within about a mile from the town, well deserves a visit from those who are fond of exploring natural wonders. It is one of the more considerable of the caverns in the mountain-limestone formation. The entrance on the side of Grin-low, below the plantations, is extremely contracted; but after a few yards, it becomes more lofty, and leads to extensive chambers, through the bottom of which a narrow streamlet channels its way, and over which are roofings and arches of imposing extent and character; stalactites hanging from the roof in some places, and large crystalline masses having accumulated on the flooring of the chambers in many places, from the dropping and welling of the water charged with calcareous matter.

The drives in the neighbourhood of Buxton are likewise very interesting.

There is a pleasant drive over Fairfield, and by Dove Holes and Barmoor Clough, to Chapel-en-le-Frith, and thence to Horridge End on the Manchester road, and thence to Buxton. This is a distance of about fourteen miles.

The village of Wormhill is reached by a road across the Fairfield common, somewhat to the right of the road just mentioned. The carriage may be left on the road near to Wormhill Hall, while the great rock, Chee-Tor, and the valley which it abuts upon, are being explored. Returning to the carriage, the road may be followed thence to the end of Millar's dale, opposite to Priestcliff, whence the return to Buxton may be by the road which joins the Bakewell road at the fifth milestone. This presents a very interesting circuit of about fourteen miles.

An interesting long drive, about twenty-eight miles in circuit, full of variety as to the objects and scenery, is obtained by following the Ashbourne road from Buxton, until within a few yards from Brier Low toll-gate, between two and three miles from Buxton; following the road to the right, leading through Hind Low toll-gate, with the higher upland of Brier Low on the left, the elevation of which is 1481 feet; the road passing thence under an archway of the High Peak railway, and thence, at a higher point, presenting views, on the right hand, of the extraordinarily twisted-looking, pointed, jagged, and irregularly shaped hills, Tor Rock, Swallow Tor, and Chrome Hill; and, as the road descends rapidly to the hamlet of Glutton, leaving Earl Sterndale and High Wheeldon on the left. Near to Glutton, the road passes through a bold rocky gorge: and immediately beyond Glutton toll-gate, the river Dove, here a stream of inconsiderable size, is crossed by a bridge. The river, at this place, passes near to the edge of the gritstone formation, and separates Derbyshire from Staffordshire. The road continues on the gritstone to Longnor, and thence to Warslow. The road ascends rapidly from Glutton to Longnor, and affords a fine view, on the left hand, of the

valley of Aldery, Glutton, or Crowdecote, variously so called,
with its gritstone bottoms, and its limestone hills on the
north and east, and gritstone hills on the south and west.
Near to Longnor, the valley of Hollins-Clough is seen
from the road, backed by the swelling hill sides, and sharply
defined hill-tops, of this part of Staffordshire. Having passed
through Longnor, the road is continued, through interesting
and various scenery, to Warslow, at which place the mountain limestone formation is again met with. Immediately
beyond Warslow, is the Ecton mining district, of course on
the limestone formation; the hill-sides showing indications
of the extent and importance of the mining operations
which were at one time carried on there. The Ecton mines
yielded, about half a century ago, great quantities of copper;
and they are still worked on a small scale. Good specimens
of the ores of lead, copper, and zinc, are readily found.
Immediately beyond Ecton, the river Manifold, an important tributary of the river Dove, which arises near to Longnor, and falls into the Dove at Ilam, becomes a stream of
some importance, and the course of the river becomes
exceedingly beautiful. From Ecton to Wetton, a distance
of about three miles, the road near to the river side, although
occasionally somewhat rough, is to be followed; and the
clear and bright stream, and the meadowed banks, and bold
valley sides, offer, along the winding course of the river, some
beautiful scenery. On the right, the entrance to a remarkable cavern, on the side of Wetton Low, is seen from some
considerable distance. The effect of the entrance to this

cavern, as seen from the road, is marred by the regularity of
its arch, which is often supposed to have been either formed
or modified by art, and might be mistaken very readily for a
much misplaced work of masonry. From the part of the
road which is nearly opposite to it, the cavern is readily
reached on foot. The arch of the entrance is forty feet in
width, and it is probably about sixty feet in height. The
entrance is brilliantly and effectively lighted to a considerable
depth, owing to a second entrance on the right, almost as
lofty as the principal entrance, but much narrower; and
almost opposite to this, there is a column of bold massive-
ness, supporting arches which extend further inward; and
the effect of the light, and of the size and proportion of the
arches, on returning to the entrance, is very beautiful. On
leaving the cavern, the Low should be climbed for the sake
of the completely panoramic view from the top, which can
have few equals. The borders of Wales are believed to be
visible on one side, the range of country between Ashbourne
and Derby is seen on another side, and the hills near Longnor on another, and the more immediate surroundings are
various and beautiful. A walk of about a mile leads thence
to the village of Wetton, where the carriage may be conveniently rejoined; and a drive of about four miles, through
interesting scenery, leads from Wetton to Beresford and
Hartington; and Beresford-dale should be seen, if time
permit. The return to Buxton, notwithstanding somewhat
rough roads, should be through Sheen to Longnor, over an
upland road, whence there is an extensive view of the

scenery that has been passed through, and from which Longnor, and the whole of the valley of Crowdecote, with their mountainous surroundings, may be commandingly viewed.

Another circuit, of about fourteen miles, is obtained by again following the Ashbourne road nearly as far as the first toll-bar, thence turning to the right, passing through Earl Sterndale, and over the road which skirts the eastern side of High-Wheeldon. The carriage should be left at this point, and High-Wheeldon climbed, for the sake of seeing the extensive view from the summit of its great and curious conical elevation; and then, having returned to the carriage, the journey is continued, by Crowdecote, to Longnor, returning to Buxton by Glutton-bridge, regaining the Ashbourne road at the same point at which it had been left.

The first six miles of the road to Leek present a hilly road and wild scenery, but with extensive and various views of much boldness and interest.

Six miles on the road to Macclesfield, near to the road side inn, the " Cat and Fiddle," is the very remarkable view already mentioned over Lancashire and Cheshire, which probably extends over a distance of fifty miles.

A very beautiful long drive is obtained, by leaving the Bakewell road, near to the fifth milestone,—proceeding thence by the road to Millar's dale,—keeping on the lower road near to the river, as far as Litton mill,—passing thence over Cram-side to Cressbrook,— and thence into Monsal-dale ;— and, at the part where Monsal-dale bends suddenly to the west, either remaining with the carriage and leaving the valley scenery, and proceeding in the carriage to Ashford, and thence to Buxton,—or leaving the carriage at the bend of the valley, and crossing the river, to ramble through the remainder of Monsal-dale, and regain the Bakewell road, and rejoin the carriage after it has made the circuit by Ashford, opposite to the eighth milestone from Buxton.

Bakewell, Chatsworth, Haddon-Hall, Matlock, Tideswell, Middleton-dale, Castleton, Beresford-dale, and Dove-dale, form severally, as will have been already inferred, most interesting objects for excursions from Buxton; and the most distant of them can hardly be said to be beyond the reach of being seen within the limits of a summer's day. There is a remarkably fine old church at Tideswell, which well deserves to be seen; and Middleton-dale, in the same direction, about twelve miles from Buxton, is a limestone valley of considerable beauty.

Sheffield is twenty-six, Chesterfield twenty-three, Nottingham thirty-five, and Ashbourne twenty miles from Buxton.

The mountain-limestone formation, to which this district owes so much of its character, is of considerable extent. Its greatest length, from the Blue John mine, near Castleton, which is on its extreme margin on the north, to Sally Moor (near Wootton, about two miles to the north of Alton Towers), the extreme margin on the south, is nearly twenty-three miles. Its greatest width is from twelve to fourteen miles; the narrowest part, from Middleton-by-Youlgreave to Hartington, is five miles across. These measurements are given geographically, without reference to the windings of roads. The average addition of one-sixth, will give a sufficiently near approximation to the distances by the roads.

Buxton is on the north-western margin of the limestone formation; and its boundaries may be readily traced from thence. Fairfield and its common are likewise on the margin of this formation; the higher grounds of Corbar and Black Edge being on the gritstone. The Water Swallows, at Dove Holes, are on the edge of the limestone. The road from thence to Castleton passes very near to the margin of this formation; the Blue John mine, to the right of the road, at the upper end of Hope dale, as has been said, being on the limestone formation; and Mam Tor, on the immediate left, and the whole of the districts of Chapel-en-le-Frith, Glossop, Kinderscout, and Edale, being beyond this formation. The town of Castleton is beyond the edge of the limestone; and consequently, the great cavern of the Peak is on its very margin. Middleton dale is near to its margin; Eyam, Stony Middleton, Baslow, and Bubnell, are beyond its margin. Bakewell and Ashford are on the limestone, but close to its margin. The Vale of Haddon, Rowsley, Edensor, and Chatsworth, are beyond the limestone margin. Youlgreave and the adjacent hamlets of Middleton, Elton, and Winster, are on the margin of the limestone. The village of Matlock, is on the adjoining formation; Matlock Bath and its valley are on the limestone formation, but close to its margin. The villages of

Brassington and Tissington are on the edge of the lime-stone formation. The considerable hill, Thorpe Cloud, at the southern extremity of Dove dale, is on the edge of the limestone. Ilam Hall is on the margin of this formation.

The limestone formation is surrounded by the great formation of millstone grit, interrupted occasionally by limestone shale; and this formation surrounds the lime-stone, in almost all directions, with ground of higher elevation than its own high level; the difference sometimes amounting to between eight and nine hundred feet. The gritstone formation is particularly extensive on the north and west sides, but gives a boundary to the limestone formation of several miles in breadth on the east. On the south, there is a very narrow and irregular edging of gritstone, which separates the limestone from the new red sandstone formation.

The climate of Buxton is necessarily much affected by the physical conditions, which have now been shown to obtain throughout the extensive tract of elevated country on which it is situated; and which surrounds it to an average distance of from twelve to twenty miles, in all directions.

The elevation of the lowest part of Buxton may be said to be 1,000 feet above the level of the sea; the level of the new church having been ascertained to be 1,029½ feet. The mean density of the air is by so much less than it is in most places. In the words of Mr. Whitehurst, "the column of quicksilver in the barometer-tube is always an

inch lower at Buxton than at Derby, at the same time and
under similar circumstances." This difference in the degree
of atmospheric pressure has much effect on the human
system; and renders the removal of invalids to this district
from places that are situated at a lower level, either eminently advisable or otherwise, according to the nature or the
stage of the ailment. The effect of a change to this
mountain air, on comparatively healthy people, is exciting
and invigorating,—promoting circulation, appetite, and
digestion, and increasing the buoyancy of the feelings and
the general energies of the system. In the instance of the
invalided, the probability of such excitement, the degree of
which may be increased by the mobility which so often
attends severe or long-continued morbid action, may render
inadvisable the removal to such an air as that of Buxton.
This observation is more especially likely to be applicable to
those, who are suffering from more acute morbid states; and
especially when attended by much mobility of circulation,
and great susceptibility of tissue. On the other hand, the
mere debility and relaxation, which are so often consequent
upon the removal or mitigation of acute disorders; and the
mixture of debility and congestive torpidity, which so often
accompanies protracted convalescence, and so often attends
protracted indisposition of chronic character, are much
relieved by removal to the thinner and less oppressive air of
this elevated district.

The physical character of this mountain country may
help to explain the comparative immunity, which is enjoyed

by its inhabitants, from epidemic and endemic diseases. The dry and absorbent soil, which distinguishes almost the whole of the limestone formation, and the greater part of the gritstone formation,—assisted, as this is, by their elevated position,—must conduce much to this result, by diminishing the amount of stagnant water, and other sources of miasmatous impurity. Even the ordinary exanthematous epidemics (measles, scarlatina, and the like) are usually of singularly mild character in this district; and typhus, and even common continued fever, rarely occurs, unless when brought into the district by persons who have been sojourning in less favoured places; and when thus met with, has very seldom been known to have been extended to a second case. It has to be said, moreover, that no case of epidemic cholera has occurred in Buxton during any of the visitations of this fearful disease.

The bearing which the high elevation of Buxton above the level of the sea, appears to have upon the probability of its exemption from cholera, is very interesting and curious. It was announced by Mr. Farr, in the admirable report on cholera for 1849, and has been since confirmed by further statistical experience, that the mortality from this dreadful malady increases as the level of places in an affected district is lower; and not only that localities of higher level have a less liability to the disease, but that the degree of immunity may be measured by the degree of the elevation. This law must be liable to contradiction or disturbance under very powerful local exciting and predisposing circumstances, and

would not be held to justify in any case the neglect of the
common principles and conditions of public hygiene ;
but it is curious as a great truth, derived from extensive statistical data, and interesting in regard to
the climatorial and hygienic character of the Buxton
district. And this may help to explain the observation,
that influenza is believed to have been a much milder
ailment in Buxton than in most other places, both as
to its immediate severity, and its ulterior consequences.
The importance of this could hardly be over-stated, when it
is remembered how much more frequently the visitations of
this great catarrhal epidemic have occurred of late years ; and
that they are believed, either directly or indirectly, to have
resulted in a greater amount of fatality than any other
epidemic disease. In forming an estimate of the degree of
fatality which attends influenza, "It is necessary," as
Dr. Theophilus Thompson well says, in the able volume
compiled by him for the Sydenham Society, entitled
" Annals of Influenza," " to extend our consideration to the
fact that during the prevalence of epidemic catarrhal fever,
the mortality is usually increased, often to a very remarkable
degree. The cause of influenza, independently of its agency
in producing characteristic symptoms, appearing to exert a
power to modify any pre-existing disease with which it may
combine, to impair extensively the vital energy so as to
increase in the population of an infected district, the
liability to contract other diseases, and also to lessen the
ability to resist any degree of fatal tendency which such

E

concurrent diseases may possess." Such is, beyond all question, the explanation of much of the indirect fatality, which is referable to all the more severe epidemic diseases. To form a true estimate of their effects, the mortality that can be directly estimated, must be considerably added to. And when it is considered that, according to the able and trustworthy reports of the Registrar-General, nearly one fifth of the total mortality of England is referred to "epidemic, endemic, and contagious diseases," too much importance can hardly be attached to the physical and sanatory character of any locality, in which a considerable degree of exemption may be obtained from these fatally influential classes of disease. If the mortality from all such diseases may be estimated, with much probability, to be in the ratio of one in twenty of those who are attacked, the smaller amount of sickness, protracted indisposition, and resulting debility, which a large degree of exemption secures to the inhabitants of this district, in addition to the lower rate of the probable mortality, deserves to be prominently mentioned in this work. It cannot be disputed, that a remarkably high average of health is enjoyed by the inhabitants of this district; the generally healthy aspects of the children are the subject of frequent observation, and the large number of people who live to an advanced age has always been noticed. Such popular statistics, however, if unfounded on precise data, are not to be received with implicit trust; but that there is an important degree of exemption from endemic and epidemic diseases, that they are generally of comparatively mild character when they do occur, and that all disease throughout the district is commonly of very mild and simple type, is the universal experience of the medical residents.

The mountain position of Buxton and the surrounding districts, renders the whole locality colder than lower situations in the same latitude. The inland position, moreover, renders the mean annual temperature less equable, than in places which are more within the influences of the oceanic temperature. There are two circumstances which deduct considerably from these disadvantages. The one of these is the degree of shelter from prevailing winds, which is afforded by surrounding grounds which have a still higher elevation. There is an appreciable difference of temperature between the upper and lower parts of Buxton in cold weather, and more especially if accompanied by high winds, blowing from the north or north-east; and there is a much greater difference of temperature between Buxton and the surrounding villages, which are less sheltered. But within these forty years, many hundreds of acres have been covered with plantations, now of an advanced and important growth; which serve, not only to clothe and embellish the scenery of this large-featured country, but to shelter it from the winds in an important degree. The second modifying circumstance referred to, is the relative dryness of the air of Buxton. The comparative essential dryness of the air of a place, greatly subtracts from the effects of an absolute lowness of temperature. Dr. Kilgour says that "cold

moist air, compared with cold dry air, abstracts caloric from
the body in the ratio of three hundred and thirty to eighty
degrees," or more than four times as rapidly. In other
words, the chilling effect of cold moist air is four times
greater than that of cold dry air; and therefore the absolute
lowness of temperature at Buxton, compared with the temperature of less elevated places, may be sometimes fully, if not
more than compensated, by the dryness of its atmosphere.
The dry air of the Buxton district is one of the most interesting of its characteristics: it is in general singularly free
from fogs and exhalations, and remarkably clear; enabling
the objects of distant scenery to be seen with most defined
distinctness. This is partly due to the altitude of the place;
partly, to the attraction of the clouds and condensed vapours
in the higher regions of the air, to the more elevated grounds
in the neighbourhood; and partly, to the comparatively
small amount of aqueous exhalation, charged with organic
contamination, owing to the absorbent nature of the limestone and gritstone soils. It should likewise be said, that
immediately after the heaviest rains, the grounds and walks
and roads in the town and neighbourhood, are found to be dry
to a degree which excites the surprise of all strangers. And,
therefore, although, as in most mountain districts, more rain
falls in the High Peak of Derbyshire than in most places of
less elevation, the effect of the rain is less observable or
inconvenient. But the Buxton district does not receive its
due share of all the rain, which its mountains may be the
means of collecting. It is often observable, that clouds

which are collected around Axe Edge, are precipitated on
the adjacent districts of Cheshire and Lancashire, when
there is no rain near Buxton. The enclosure of thousands
of acres, which were unenclosed and bare of pasture within
the memory of the present generation, and the drainage of
extensive districts for agricultural purposes, are pro-
gressively influencing the climate and scenic character of
these upland districts. It is almost difficult to imagine
how recently a large proportion of the country around
Buxton, and within many miles of it on all sides, was
without an enclosure, and covered with heath, gorse,
and rank vegetation, the trees having been as yet un-
planted which now so much embellish the landscape in all
directions.

The beneficial effects of the air of Buxton upon some of
the invalids who resort to it, cannot be wholly explained
even on the ground of its mountain elevation, or on that of
its dryness. It sometimes happens, that invalids from the
neighbouring gritstone districts, of an elevation at least as
considerable as that of Buxton, are more benefited, and
more rapidly benefited, by removal to the limestone formation, than the mere change of air, or any other concomitant
circumstance, can serve to explain. There is indeed usually
an amount of stimulating effect, produced by the limestone
atmosphere upon those not accustomed to it, which neither
the elevation of the place, nor the dryness of the air, can be
held to explain. It frequently happens, that invalids, who
are strangers to the place and district, and who state that

the remark has not been derived from any second person's suggestion, affirm explicitly that they can *smell* the air of the limestone. On enquiry, it has seemed, that they have experienced a tingling sensation in the nostrils from the air, rather than that the air has been really odorous to them. That there may be important atmospheric differences, which are not appreciable by chemical re-agents, is universally admitted; and it is not too much to suppose, that the air may be influenced by passing over or resting upon extensive districts, either according to their vegetative surface, or their geological character. There is some such effect produced on the air of this district, by the great limestone formation; and the result of this is, more particularly, to add to its stimulating influence on the human system.

The spring in Buxton is unusually late, and proportionably short; the summer is of the average duration; and the autumn is long. The spring can seldom be said to have crept from the arms of winter until the month of April; and, in general, April is near its close before the winter can be said to be fairly got rid of. From the middle of May to the end of October; and in some seasons until considerably later, and almost to the end of the year; may be said to constitute the real spring, summer, and autumn, of Buxton. July and September are apt to be wet months throughout England. The latter end of May, the whole of June, of August, and of October, are usually the least changeable periods of the year.

The Buxton waters, whether used as baths or internally,

are equally efficacious at all periods of the year; * and those
who are suffering severely from those ailments in the relief
of which they act so powerfully, make use of them at any
time throughout the year. But people who have a choice,
generally prefer the summer and autumnal months for
migrating to watering places; for the obvious reasons, that
the country always looks best when nature has donned her
livery, when the sun is bright, and the weather warm; and
that exercise, which is so valuable an auxiliary to all
medicinal treatment in chronic cases, can then be taken
most pleasantly, and perhaps with most advantage. It is
difficult, however, to understand, why Buxton should not be
as full of its invalided visitors in June and July, as it is in
August and September. This may have arisen from fashion
and secondary circumstances; but time and common sense
must one day show its absurdity. The comparatively cold
evenings and mornings of autumn, and the greatly shortened
days, render the exercise which can be taken at this time
of the year, less continuous than that which may be taken
in the later spring months and the summer months, and make
the disposal of his time irksome to many an invalid, in whom
indisposition may have spoiled the taste for reading, and
for whom the excitement of much society may be necessarily

* "The usual season for drinking the waters is from the beginning of
May, to the latter end of October; but if the patient requires a longer
perseverance, he may safely use them all the winter, as they are found,
upon repeated trials, to be equally good in all seasons."—Dr. Hunter's
"Buxton Manual, or Treatise, on the Nature and Virtues of the Waters of
Buxton." York, 1765.

disadvantageous. When invalids visit Buxton during the
months of June or July, the days being then long, and dusk
and bed-time but little divided from one another, the
cheerfulness is more likely to be maintained, and the
spirits and feelings made conducive to the effects of the
mineral waters, rather than suffered to intefere with them;
for it need not be said, that the influence of the mind upon
the body, at all times paramount, is more especially
important for good or for evil when the latter is affected
with disease.

I have used the word "continuous" with reference to
exercise; and it may be well to embrace this opportunity of
expressing my opinion, that every one, and invalids especially,
should take the quantum of exercise, not only regularly,
but in divided doses, at different times of the day, rather
than endeavour to do as much as can be done at one time,
by which fatigue is induced, the blood is determined almost
unduly to the surface of the body, the internal organs are
disturbed, and the nervous energies are inconveniently, and
it may be injuriously, expended. This deserves to be
seriously considered in regard to invalids, who may be enfeebled by indisposition, and whose systems may be predis-
posed to be acted upon by causes apparently the most
unimportant. On such persons it cannot be too strongly
urged, that they should not attempt to walk far at one time;
but rather, that a number of short walks should be alter-
nated by rests. This kind of exercise, which may be taken
within an hour or two after breakfast, and continued for

longer or shorter times, with longer or shorter intervals, almost until bed-time in the summer months, is that which will be found to be most useful to nearly all invalids, which will employ the mind most fully throughout the day, lessening the chance of the time seeming to be hanging heavily on their hands ; while a more constant exposure to the genial and tonic influence of the air will be promoted : a point which, in regard to the value of Buxton as a watering-place, deserves perhaps a higher degree of importance than has as yet been assigned to it.

But there is even a stronger reason to be advanced, why invalids should resort to Buxton as soon as the return of warmer and more settled weather permits them to leave home with comfort. There is more general activity in the system, at this time of the year, than at any other season ; and a correspondingly greater natural effort made for the relief of the more chronic ailments, than at any other period. This will have been noticed by most persons, as well as by medical men. The return of spring, of warmer weather and brighter days, not only stimulates the dormant vege-tation, and arouses the life of plants into a renewal of activity ; but acts upon animal life likewise, and includes man in its effects ; notwithstanding the artificial condition, and the artificial wants, which civilization has given rise to. The effect of such stimulus on the human system is seen, in a very obvious and painful degree, in the instance of those who are suffering the extreme stages of disease. The warmer air is too stimulating for the irritated and wasted

tissues ; and it has to be predicated of many such cases, that
the more early the return of the season which is more
genial to the healthy, the sooner will the slender thread
be divided by which such invalids cling to life.   And it is
no less observable in other cases, in which the effect of such
stimulus is not to destroy life, but to aid in the restoration
of strength, and the removal of disease.   It has often been
suggested, in the instance of sufferers from the most chronic
and obstinate of the disordered conditions, in the relief of
which the Buxton waters are found to be so generally
useful, that, in order to obtain the greatest amount of effect
from their use, the earlier periods of the year should be taken
advantage of for their use, when the efforts of the system,
thus stimulated by the season, will be most likely to aid and
confirm their effects.   This stimulating effect of spring upon
the human system may be increased by the periodical type,
which all the operations of nature are found to assume.
There is a remarkable tendency to periodicity in all vital
phenomena.   It is not merely in regard to the seasons, the
succession of day and night, and the alternation of repose and
activity ; but this law extends to all natural phenomena.
And this great law may be supposed to aid in producing the
development of all the vital efforts, which is so observable
in spring, and which is found to be of so much practical
importance in regard to disease.   The dryness of the air of
Buxton is the probable cause of one singularly valuable
peculiarity, which is observable in regard to the place.
Invalids hardly ever take cold at Buxton.   They may have

only just quitted the bed-room to which they had been confined by serious and protracted indisposition, have left the comparatively close and heated air of the house, have travelled to Buxton closely muffled and packed in wrappings and multifarious envelopments ; and, almost immediately on arrival, have ventured into the open air, have sat down on the benches with which the walks are abundantly supplied, and have spent hour after hour in this indulgence, which in most places would prove to be so dangerous ; but it is here hardly ever followed by any unpleasant consequences ; and on the contrary, contributes to restore health and vigour to the enfeebled system. When it is considered, that this observation has reference to a place which is resorted to for the use of tepid and hot baths, from the effects of which the vessels of the surface would be rendered unusually susceptible, its importance in a medical account of Buxton will be admitted to be great. The statement, however, must not be made use of to justify rashness, and sudden and violent changes in regard to exposure and clothing on the part of invalids; but it shows that to be justifiable which would otherwise be unsafe ; and it illustrates strongly a valuable character of the air which superincumbs the mountain limestone.

Change of air is well known to be capable of producing much effect on the health of man. The amount of such effect is usually in proportion to the degree of the change, provided that it is not too great for the powers and susceptibilities of the system to endure without injury, and that the change is from a less pure to a purer air, and from an air that is damp to one that is dry. There are important disordered conditions in which the removal to a damper air is indicated, and in which a low and well sheltered situation should be preferred. But these cases are the exception to what applies to the majority of disordered states. It cannot be wondered at, that the removal to such a locality as Buxton should be followed by beneficial results, in many of the diseases from which the inhabitants of low and damp localities suffer so severely. It must indeed be admitted, that change of air of any kind often does good. The secondary influence of mind and its associations, may often justifiably assign a preference to such a change of air, as might seem to be less suitable to the generality of cases. There is, for instance, a virtue in the air of the native place, which may be inexplicable, unless the indirect effect of memory and association be allowed for. It has been said, that

> "Custom moulds
> To every clime the soft Promethean clay :
> And he who first the fogs of Essex breath'd,
> (So kind is native air) may in the fens
> Of Essex from inveterate ills revive,
> At pure Montpelier or Bermuda caught."
>
> ARMSTRONG.

If this were so, the said " custom " would probably have little to do with the result. The effect would have to be ascribed to the mental stimulus : to the influence of the scenes of the younger days on the mind, and through the mind on the body.

But, waving a consideration of such influence of mind, and of such special or exceptional cases as have been referred to, and of the cases in which disease may have so nearly done its worst, that an exposure to the ordinary breath of heaven could hardly be undergone with impunity, and to which the mildest and blandest air may alone be suitable, it may be affirmed that the locality, in which the air is the most dry and pure, and the inhabitants are most free from disease, will be the most useful to the generality of invalids, and that Buxton deserves to be ranked among the places which claim so high a character.

CHAPTER III.

The temperature of the Buxton tepid waters is the first
of their characteristics to be noticed. It is one of the most
important of their sensible properties; for, although no one
who had been witness to their medicinal effects could ascribe
them to the temperature of the waters, yet this temperature
must aid and increase such effects, and facilitate the admission
of the saline and gaseous constituents into the systems of
those who make use of them.

There are few subjects which have given rise to more
speculation and inquiry, than the cause of the elevated
temperature of hot springs. Although there are only two
districts in which such springs are found in England, they
are by no means rare in most other countries. More than
forty such springs are reported to exist in Portugal alone;
the temperature of which ranges from 68° to 150°.
Between sixty and seventy of these springs are said to exist
in France; the temperature ranging from 70° to 221°.
Switzerland and Italy are likewise rich in this respect.

But the springs of Germany have far outstripped all the others in medical importance ; which may be ascribed, in some degree at least, to the personal efforts of the highly gifted people of that country. The baths of Aix-la-Chapelle, Wiesbaden, Ems, Baden, and several others, are as well known by reputation in this country, as to the Germans themselves.

Before directing attention to the causes which may produce the elevated temperature of hot springs, it may be endeavoured to be shown whence the water is derived with which these springs are supplied.

The water of all springs is derived from the atmosphere, or from large subterranean reservoirs, or from the ocean.

The water of most springs is derived from the atmosphere. The aërial vapours are condensed on the surface of the earth, in the form of rain, hail, or snow. The water percolates through the softer strata, finds its way through fissures or *faults* in the denser strata, until its progress is stopped by an impermeable bed of clay, &c. It is forced over the surface of such a stratum, by the pressure of the superincumbent water ; until it is carried once again to the surface of the earth, through a breach in the stratification above it, and forms a spring. But miners say, that the lower they descend below the surface, the less water they meet with. Indeed, these waters can hardly be supposed, under ordinary circumstances, to penetrate very far, without being absorbed by the strata through which they pass, or arrested, and again brought to the surface, by meeting with an impermeable stratum. Whereas, boiling water is poured out by volcanic agency, at an elevation of many thousands of feet, on the confines of perpetual snow; and consequently, the depths at. which large collections of water may be supposed to exist, may be inferred to be very considerable.

There seem, then, to be reasons for believing, that there are reservoirs, or large collections of water, situated at very considerable depths in the bowels of the earth; while it must be added, that such collections of water can hardly be so considerable, as to be independent of supplies from other sources, and to be capable of pouring out the immense volumes of water, which are discharged from depths, to which the atmospheric waters cannot be supposed to penetrate, under ordinary circumstances; or such discharges of water, in quantities so considerable, would materially increase the quantity of water contained in the atmosphere and in the ocean. It has been inferred, that such subterranean reservoirs cannot be of such extent as to afford this large supply, from calculations by which a specific gravity is assigned to the globe of nearly five times that of water, and a third more than the mean density of its rocky crust. Such inferences, however, are without adequate foundation. The Andes, with their elevation of 25,000 feet above the level of the sea, have been well said to bear no greater proportion to the size of the earth, than the roughness on the rind of an orange to the size of the fruit; and yet, compared with this, how little is the greatest depth to which miners have explored the crust of the globe. A mine in the Tyrol is stated to be 2764 feet in depth ; and this is probably the greatest depth to which man has penetrated. We can be little justified with such facts before us, in forming conclusions as to the composition or character of the internal structure of the earth.

But when we take into account the amount of volcanic agency, which is still going on in various parts of the earth's surface ; when we consider that thermal waters spring, with few exceptions, in the neighbourhood of either recent or extinct volcanoes, or of such disrupted stratification as are the results of volcanic action ; when we couple with this, the amount of volcanic power which is manifested in the ocean, —the islands that have been upheaved from its depths, within the memory and authenticated traditions of men,— and the shocks that are often experienced far out at sea ; when we connect these facts with another singular fact, that thermal springs, and indeed that volcanoes, are very rarely found at any great distance from the sea,—and, indeed, when so found, are believed to derive their waters from great inland lakes, or inland seas ; and when to these considerations it is added, that columns of watery vapour, and showers of boiling water, are among the principal phenomena of active volcanoes, it is surely not too much to connect intimately the waters of the ocean with thermal springs, and to conceive that the ocean is the probable source from which the waters of these springs may be derived. It does not appear to be too much to conceive, that, through disrupted strata, or *faults*, at the bottom of the ocean, through chasms created by volcanic outbreaks, water would be forced by the enormous pressure of the mass of waters above; or that this should, under such pressure, and with the facilities created by the strata having been disrupted by the volcanoes, penetrate much deeper than we can conceive the atmospheric waters to penetrate. It may be conceived that, in this way, water may arrive at a centre of volcanic agency; and that there, urged by the heat that exists in the depths of the earth, this water may be converted into steam; and that, when the steam thus formed can find no vent, it may at length accumulate such power as to upheave the masses of strata above it, and in its turn become the active element of a volcano; or that, when the steam can find a vent for itself, by passing through strata already disrupted, it may be gradually condensed and cooled, until it may emerge at length on the surface of the earth, in the condition of a hot or of a tepid spring, according to the length of the channel through which it has had to pass.

This is perhaps little more than theory; but it is theory founded on such an association of facts, as seems to justify a strong opinion of its probability. If it be denied, it is difficult to assign any satisfactory reason, why thermal springs are not found as commonly in the interior of vast continents, as in the neighbourhood of the ocean; why thermal waters are so constantly found to be connected with existing or extinct volcanoes; and why water, in the form of steam, or at a much elevated temperature, is so constantly associated with every volcanic outbreak. But this supplies an explanation of several circumstances, which form singular and interesting matters in the history of thermal waters. It gives a key to the surprising fact, that thermal springs flow in unvarying quantity, and at an unvarying temperature, from age to age ;* and that, as far as can be ascertained, their constituents have been unmodified by time. If the waters with which these springs are supplied, were derived from the same source as that of the generality of springs, viz: the waters condensed on the surface of the earth from the atmosphere, they would necessarily be subjected to the same vicissitudes. In a particularly dry season, the supply would be diminished, or would temporarily cease; and on the other hand, after an unusually large fall of rain, the quantity discharged from them in a given time would be greatly increased. But in regard to thermal springs, these causes have no influence. In winter and in summer, in dry seasons, and in wet seasons, a certain number of gallons per minute are poured out with an undeviating regularity. Moreover, if these waters were

---

* "The springs in Greece still flow at the same places as in the Hellenic times : the spring of Erasinos, on the slope of the Chaon, two hours' journey to the south of Argos, was mentioned by Herodotus ; the Cassotis at Delphi, now the well of St. Nicholas, still rises on the south of the Lesche, and its waters pass under the temple of Apollo ; the Castalian fount still flows at the foot of Parnassus, and the Pirenian near Acro-Corinth ; the thermal waters of Ædepsos in Eubœa, in which Sylla bathed during the war of Mithridates, still exist. I take pleasure in citing these details, which show that, in a country subject to frequent and violent earthquakes, the relative condition of the strata, and even of those narrow fissures through which these waters find a passage, has continued unaltered during at least two thousand years."—*Cosmos* (*Lieut.-Col. Sabine's Edition*) vol. i.

derived from the atmosphere, the colder the season, the lower would be their temperature; and certainly, likewise, the larger the quantity of the water poured forth, the lower would be its temperature. But, through long series of years, not only does the quantity of water poured out by these springs in a given time, remain the same as it was at the most remote record, but their temperature remains steadily the same as when first noticed. Supposing the ocean to be the source from which these springs derive their supply of water, the definite quantity may be forced through the fissures, either by its gravity, or by the pressure of the superincumbent waters,—this definite quantity may be converted into steam, by the means presently to be noticed, —the steam may be condensed, and cooled to a definite degree, by passing through a definite space,—and supposing no waste to arise, from leakage or otherwise, 129½ gallons of water per minute, containing the same gaseous and saline constituents as now and heretofore, may continue to supply the natural baths of Buxton throughout ages to come.

The absence of sea-salt in any of the thermal springs, might at first be considered to be fatal to the above views. But it has been sufficiently proved by direct experiment, that sea water is deprived of its saline constituents, by passing through a certain thickness of sand, &c.; and, therefore, the passing through we know not how many miles in depth of various strata, would necessarily deprive it of all its saline matters, even if the hypothesis of its conversion into steam, and consequent separation from every foreign ingredient, were thrown aside.

Knowing as we do the gigantic extent of the processes of nature, and the uniformity of the means used to effect the same results, in different places and at different times, it may probably be inferred that the same cause, which produces the supply of water in the instance of any one of the thermal springs, might equally serve to account for it in the others. Some general source, which may equally serve in regard to all thermal springs, is therefore sought for. But in regard to the Buxton thermal springs, the circumstance obtains that, in various situations, at no far removed distances, in the limestone formation, surface springs are swallowed up by fissures or cracks in the strata; some of which are found to re-appear at the surface, at different distances from what is locally called *the swallow*; and the whole of these may or may not so re-appear. It could not be thought to be impossible, that such swallowed up spring or springs should travel through such fissures, and serve to supply some vast subterranean reservoir, from which the supply of water to be vaporised might be obtained.

The cause of the elevated temperature of thermal springs, is a question of still more immediate interest.

As has been said, thermal springs are always found in the greatest abundance, in the neighbourhood of active or recently active volcanoes; and volcanoes are hardly ever found to exist, without giving rise to springs of tepid water. In those situations where no traces of volcanic agency can be detected in the neighbourhood of tepid springs, these waters are found to issue from the primary rocks, either directly, or from beds of inconsiderable thickness, which evidently form merely a crust over rocks of the primary class. In some instances, the tepid springs are found in the midst of chains of mountains, or close to their base; in other instances, a succession of such springs is found in the same direction as that in which a mountain chain extends; sometimes, they "gush out at or near to the line of junction between the granite or other igneous products, and the stratified rock resting upon its flanks, which from its highly inclined position would seem to have been upheaved; whilst in a few cases where they occur in the midst of the granite itself, patches of stratified rock are found contiguous. Thus the same agent which forced up the granite through the axis of the chain, may have given rise to the hot springs which accompany it first along the line of the disruption. * * * * In many instances where the *general* aspect of this country does not so forcibly impress upon the mind the idea of volcanic forces having been in active operation, there is something in the *particular* circumstances of the locality indicative of the same kind of agency." (*Professor Daubeny on Volcanoes.*)

Professor Daubeny goes on to cite the tepid spring at Clifton, as gushing out of a narrow fissure of hard rock, bounded by abrupt cliffs, with an enormous *fault* near to the mouth of the spring "which has thrown down the limestone beds one hundred and twenty feet," serving to justify the opinion that a mighty force has at some period rent these rocks in sunder, and opened a passage for the tepid water from deeply-seated strata ;—he cites Matlock " from the abruptness of the cliffs which bound the defile on either side, and from the existence of an enormous fault, much of the same description as that of Clifton," and justly adds " that the volcanic rocks which are found in many parts of Derbyshire afford an additional presumption that the tepid waters of that country owe their origin to volcanic heat ;"— he cites the warm springs of Carlsbad, as emerging from " a kind of conglomerate, composed of broken masses of granite united together by a siliceous cement," leading to the inference of riven rocks and shattered fragments and disrupted strata ;—he cites the warm springs of Pfeffers, in the Grisons, bursting forth from the side of an extraordinary chasm in a limestone rock ; adding that " the other thermal springs in Switzerland appear under circumstances for the most part similar," and that "the situation of the thermal waters in the beautiful mountain region of Virginia, west of the Blue Ridge, which I visited in 1838, strongly corroborates the views above enunciated ;" "in short, out of fifty-six springs more or less thermal, forty-six are situated on, or adjacent to, anticlinal axes ; seven on or near lines of fault and inversion ; and three, the only group of this kind yet known in Virginia, close to the point of junction of the Appalachian with the Hypogene rocks."

The position of the different tepid springs in Derbyshire confirms these views strongly. Not only do the broken and shattered strata, and the abrupt cliffs, and frequently occurring patches of toadstone, tell of volcanic action and riven rocks, and account for fissures by which such springs could find egress; but the springs occur near to the edge of the limestone formation in every instance; and in such situation, the continuity would be more likely to be broken through, down to more deeply seated strata.

Connecting these facts together, the conclusion seems to be justifiable, that thermal springs arise from beneath rocks of the primary class, through disruptions which have been caused by volcanic agency; and granting that thermal springs arise from terrestrial depths below all the strata which have been the subject of geological knowledge, it would remain to be shown whether the temperature of the interior of the earth may be adequate to raise large quantities of water, brought into successive contact with it, to the boiling temperature.

It has been observed, in many countries of high latitude, that when the atmospheric temperature falls below a certain point, the temperature of the springs in those countries ceases to fall in the same ratio; and, in fact, that their temperature often exceeds that of the air. Nor have these singular observations been confined to the springs of the countries referred to. It is well known, that a certain elevation of temperature is essential to the life of plants, and that different plants have different ranges of temperature within which they can live. It is said to have been ascertained, that rye requires for its growth a temperature

of not less than 46°; and yet, owing to the internal
temperature which emanates from the earth, independent of
the solar influence, this grain is grown and ripened in
Sweden, where the atmospheric temperature is little more
than 36°. It would seem indeed, that the mean terrestrial
temperature exceeds the atmospherical in many northern
districts, if not in northern countries generally; and that it
is owing to this, that nearly the whole of Siberia, the upper
parts of Finland, and some parts of Sweden, afford harvests
and sustenance to the inhabitants, under a degree of
atmospheric temperature which would be insufficient for
these purposes, but for the inherent temperature of the
earth. That this cannot be owing to the absorption of the
solar heat during the warmer months, appears to have been
proved by experiments, showing that six months are required
for the absorption of heat to the comparatively trifling depth
of thirty feet. But this is proved even still more conclusively
by the ascertained fact, that the atmospheric temperature at
the equator is higher than that of the perennial springs.

These observations would go far to establish an opinion,
that the earth possesses a considerable degree of internal
heat, which would almost necessarily become more considerable the greater the distance from the surface. But it is
by the results of experiments which have been made in
mines, that it has been established as a fact, that the deeper
we penetrate beneath the surface of the earth, the higher
the temperature is, and that we are enabled to form some
idea of the depth at which the earth may be at so high a

F

temperature, as would suffice for the conversion of water into steam.

In the ancient quarries below the observatory at Paris, at the depth of only ninety-two feet, the temperature is nearly 2° higher than that of the mean temperature of the country. If the temperature of subterranean springs be taken as a guide, to indicate the increased temperature of the earth as we penetrate more deeply below its surface, it has been found, to cite one out of many such observations, that in the copper mine of Dolcoath, in Cornwall, at the depth of 1440 feet, the temperature of the spring is 82°, while the mean temperature of the country is only 50°; or, to mention another instance, that in the silver mine of Guanaxuato, in Mexico, at the depth of 1713 feet, the temperature of the springs is more than 98°, and the mean temperature of the country little more than 60°.

But it is chiefly by ascertaining the temperature of the rock itself, at different depths, that a fixed conclusion may be arrived at, as to the rate of increase in the temperature, as we descend more and more deeply into the bowels of the earth. And the result of such observations, many of which have been made, with great care, and possible sources of fallacy watchfully guarded against, is that the earth becomes warmer by one degree for every forty-four feet of depth; and, consequently, at a depth of little more than 7000 feet below the surface of the earth, the temperature would be sufficient to raise water to the boiling point, and convert it into steam: a depth which bears no greater proportion to

the diameter of the globe than a few inches bear to
a mile.*

It must be admitted that this would be sufficient to
account for the elevated temperature of thermal waters,
were their temperature the only particular which distinguishes them from other water. The intimate connection that there is between these waters and volcanoes has,
however, led to repeated suggestions, that these may or must
have something more to do with the production of these
waters, than merely the having forced the channel by which
they escape to the surface. Snow, to the depth of two feet
and a half, remained unmelted on Vesuvius, after the
eruption had lasted two days, in the year 1822 ; and the

---

* "Hot springs," writes Baron Humboldt, "issue from rocks of every
kind ; the hottest permanent springs yet known are those found by myself,
at a distance from any volcano,—the 'Aquas calientes de las Trincheras,'
in South America, between Porto Cabello and New Valencia, and the
' Aquas de Comangillas,' in the Mexican territory, near Guanaxuato. The
first of these had a temperature of 194·5° Fahr., and issued in granite ;
the latter in basalt, with a temperature of 205·5° Fahr. According to our
present knowledge of the increase of heat at increasing depths, the strata,
by contact with which these temperatures were acquired, are probably
situated at a depth of about 7800 English feet, or above two geographical
miles. . . . . . The elevation of the new volcano of Jorullo, unknown
before my American journey, offers a remarkable example of ordinary rain-
water sinking to a great depth, where it acquires heat, and afterwards reappears
at the surface as a thermal spring. When, in September, 1759, Jorullo
was suddenly elevated to a height of 1682 English feet above the surrounding
plain, the two small streams called Rio de Cutimba and Rio de San Pedro
disappeared, and some time afterwards broke forth afresh from the ground
during severe earthquake shocks, forming springs, whose temperature, in
1803, I found to be 150·4° Fahr."—*Cosmos.*

observers were able to keep their naked hands on the
margin of the lava stream without inconvenience, at a time
when the centre of it was still in a fluid state. This proves
how slowly heat passes through the volcanic products; and
it has been urged that it is possible there may be masses of
melted material, thus crusted over, of enormous size,
situated at great depths in the bowels of the earth; and
that such masses may have retained a highly elevated
temperature, during periods long anterior to any of our
records; and that currents of water, passing close to, or
near, these masses, would be vaporised by them, and might
form hot springs, the temperature of which might not
necessarily undergo any perceptible diminution during
hundreds of years.

With satisfactory proof of the astonishing fact, that at a
few thousands of feet below the earth's surface, its strata
are at a great elevation of temperature, it would seem
needless to indulge in speculation as to any other cause
for the heat of waters, which are known to proceed from
great depths, and probably from greater depths than geology
has made us acquainted with, and at which no other cause
than the temperature of the globe itself would be needful
to convert water into steam.*

---

* "The relation, indeed, of almost all springs impregnated copiously
with mineral matter to the sources of subterranean heat, seems placed
beyond all doubt by modern research.. Mineral waters, as they have been
termed, are most abundant in regions of active volcanoes, or where earthquakes are most frequent and violent. Their temperature is often very

· But, in truth, the ingredients of mineral waters, both gaseous and saline, being identically the same as the materials discharged from the bowels of the earth in volcanic eruptions, it is impossible to avoid the conclusion that these waters proceed from volcanic centres, and probably derive from volcanism all their characteristics. It should be remembered that we are not driven to this conclusion, on account of any difficulty in explaining the elevated temperature of thermal waters, which might be due to the internal heat of the depths of the earth, but on account of the chemical characters and distinctions of all thermal springs. Chlorine, chiefly in combination with hydrogen, as muriatic acid,—sulphur, in combination with oxygen, or with hydrogen,—carbonic acid,—the chlorides of soda and of lime, sulphate of lime,—and oxydes of iron variously combined with carbonic acid, &c.,—are all the common products of volcanoes, and the ingredients most

high, and has been known to be permanently heightened or lowered by the shock of an earthquake. The volume of water also given out has been sometimes affected by the same cause. With the exception of silica, the minerals entering most abundantly into *thermal* waters do not seem to differ from those in cold springs. There is, moreover, a striking analogy between the earthy matters evolved in a gaseous state by volcanoes, and those wherewith the springs in the same region are impregnated ; and when we proceed from the site of active to that of extinct volcanoes, we find the latter abounding in precisely the same kind of springs. Where thermal and mineral waters occur far from active or extinct volcanoes, some great internal derangement in the strata almost invariably marks the site to have been, at some period, however remote, the theatre of violent earthquakes."—*Lyell's Geology.*

commonly found in mineral waters. And indeed all thermal waters may thus be grouped into one great family, probably identical in origin, singularly alike in chemical characters, and entirely independent of local causes, in regard to their temperature, their flow of water, and the amount of their saline and gaseous impregnation. Whereas the other kinds of springs, even although their saline ingredients may be similar to those which are contained in some of the thermal waters, are dependent on local influences, are affected by wet or by drought, and by variations in the local temperature; and their geological and geographical positions are materially different from those of the thermal springs. It seems to be an unavoidable conclusion, that the cause of volcanic action, so uncertain as to time, degree, and duration, must be chemical. The heat, the steam, the evolution of gases, all denote the operation of chemical affinities, new combinations, disturbed forces, produced and operating on a gigantic scale, and with proportionate and vast consequences. It seems to be only necessary to infer that such chemical action should be moderated in degree, by dilution of the re-agents, or by the constraining influence of mechanical difficulties or hindrances, in order to explain a more regular series of similar phenomena of less violent character; or such moderated action might follow, and be continued for long periods of time, after the chemical changes had been accomplished in regard to the substances of more powerful affinities, or presenting greater mechanical facility for their display. It may well be, that such gradations of action are equal to produce columns of boiling water or seas of melted lava, to eject them with enormous force, or fearful violence, amid the cold desolation of Iceland, or from the summit of Vesuvius,—or to charge so much water with gaseous and saline constituents, and produce a certain elevation of its temperature, and cause it to be poured forth as a thermal spring.

Theories have been broached, and arguments and facts advanced, to extend the chemical theory of volcanic action, far beyond what has now been stated. The merit of much scientific tact, and of having collected and arranged facts and observations, of great interest and value in regard to this question, is due to Professor Daubeny, whose great work on volcanoes has been already referred to.

CHAPTER IV.

GENERAL PROPERTIES OF THE BUXTON TEPID WATERS.
RESULTS OF SUCCESSIVE ANALYSIS. COMMENTARY
ON THEIR COMPOSITION, IN REFERENCE TO THEIR
MEDICINAL EFFECTS.

THE tepid mineral waters of Buxton are bright and clear in
a remarkable degree. When seen in a glass vessel, as
dispensed to the drinkers of the waters, or when seen in the
flows and conduits, their brilliancy is very noticeable. There
is, moreover, a perceptible shade of colour in the waters—a
faint tinge of blue—which is peculiar to them, and serves to
distinguish them from the ordinary waters of the district. It
is strange that this tinge of colour should have been denied
by Dr. Pearson; but he did not enjoy the same opportunity
of observing it which we now possess. In the old and
imperfectly lighted baths, the great transparency of the
waters could be noticed, but their brilliancy was wholly
unobservable, and this tinge of colour was necessarily not
noticeable. In the well-lighted apartments which now
contain the baths, the mass and depth of the water exhibit
the transparency, brilliancy, and colour, in the best
manner. The colour has been supposed, at different times,

to be due to various mineral ingredients, believed to be
contained in the waters, but not in sufficient quantity to be
detected.   At an early period, copper was thought to be the
colouring ingredient; in more recent times, iodine was
inferred to be a probable cause of the colour.   The only
purpose served, has been to indicate that the colour has
always been a subject of remark, even in remote times.
Attention has been directed to all the different causes to
which it might be referable; and unless it could be due to
the small proportion of iron, which has now been ascertained to be among its constituents, the cause of the blue
tinge is still unknown.   The brilliancy is due to the large
quantity of gas which it holds in solution, and which is
given off from it in the form of minute bubbles.   If a bottle
of transparent glass is filled with the water, and held
between the eye and the light, the water will be seen to be
charged with these bubbles; most of them being exceedingly
small, but clustered quite closely together.   In the largest of
the natural baths, at the instant of the escape of the water
from the bowels of the earth, and from the great pressure to
which it must there be subjected, much larger bubbles of gas
are given off, somewhat irregularly; sometimes as large as a
billiard-ball, and sometimes in considerable numbers.   The gas
forming these large bubbles, which are so much more noticeable than the minute bubbles, bears a very small proportion
to the quantity of gas which is given off more slowly, in the
form of the small bubbles.   The greater proportion of the gas
quickly escapes, when the water is exposed to the air; but

if carefully bottled, corked, and sealed, it would contain
much of the gas for an indefinite time. The appearance
of the large bubbles of the gas, rising like soap-bubbles
through the masses of the water in the bath, is curious and
beautiful. The quantity of gas with which the water is
charged, giving it much the appearance of the effervescing
wines, or of the artificially aërated waters after the first
violence of their effervescence has subsided, adds much to
the buoyancy of the water in the baths. Feeble invalids
have to be cautioned as to the buoyant character of the
water, as it renders care, and the having the hand-rails or
the bath-chains in ready grasp, to be needful in the instance
of infirm and feeble bathers. When limbs are more or less
paralysed, or even much enfeebled, there is sometimes a
difficulty in keeping them under the surface of the water.
In extreme cases of diminished command over the limbs, or
of great debility, the affected parts, and sometimes even the
whole body, have to be held by attendants under the surface
of the water. Unless in such extreme conditions of system
the buoyancy of the water in the baths is simply an enjoyable
characteristic of the water, and greatly facilitates the use of
muscular exercise during the immersion.

The effect of the temperature of the natural water on the
bathers, is somewhat different from what might have been
expected. The degree of shock commonly experienced at
the instant of immersion, is usually greater than would be
looked for, from bathing in water of so much higher a
temperature than that of the air. When the bath is made

use of under proper circumstances, the shock and sense of chill are only momentary effects, being immediately followed by efficient reaction. If judiciously used, there ought to be no return of chill during the stay of the bather in the water; nor should there be any chilliness, unless for a single moment, on leaving the bath. The reaction should be maintained during the whole of the time that the bather remains in the water, after the primary and momentary shock; and it should be maintained, and indeed it is desirable that it should increase, for some hours after leaving the bath. The reaction often continues throughout the whole remainder of the day, sometimes lasts throughout the following night, and is occasionally found to be continued during the whole of the subsequent day. These latter cases are rare; and, it may be observed, these extreme effects usually show that the warmer baths of these waters would be more suitable for such cases than the natural baths; and that baths of these waters, of any temperature, should be used with much caution and moderation as to the time of remaining in the water, and as to the frequency of using the bath. Such cases are, although comparatively rare, sufficiently numerous to form an important feature in the history of the effects of these waters upon the human system. And, in regard to the cases in which the shock occasioned by the bath is not merely momentary, but continues more or less during the time of immersion, or even afterwards, it may likewise be said that the natural bath is seldom used judiciously, or even without risk, under these circumstances. The use of the warmer baths will commonly be found to be preferable in these cases. The stimulating effect of these baths is not only observable in regard to the degree of reaction, or glow of increased warmth; and, in the exceptional cases, by the feverish heat which follows their use; but also in regard to the system generally. The spirits, the digestion, and the appetite, are all so much stimulated, as to convince those bathers who may have been previously the most sceptical, as to the powerful and extraordinary action of the baths of these waters upon the animal economy. In the course of the last eighteen years, a very large number of medical men have been led to make trial of the Buxton baths for the relief of their own ailments; and many of these gentlemen have had their trust in the medicinal value of the water much shaken by the older analyses, and have expressed their incredulity, in sufficiently unqualified terms, before using the bath. But this want of belief in the peculiar and remarkable character of the waters has never failed to be removed by the use of the bath, even on the first time of bathing. The gaseous character of the waters, so evident when bathing in them,—and the marked degree of excitement which follows the use of the bath,—have invariably produced a complete recantation of all preliminary doubts and disparaging opinions.

The stimulating effect produced by the bath usually lasts during the subsequent twelve or fourteen hours. In cases of disordered action, the use of the bath is sometimes followed by an increase of feverishness, pain, or stiffness, according to the nature of the ailment. This effect may begin to be felt from six to twelve hours after bathing, and may continue in greater or less degree, from a few hours to twenty-four hours, or longer. It is almost always desirable, that this effect should be allowed to subside, before the bath is made use of again. There is a more lasting effect of this kind, which frequently follows the use of several baths of this water, and which is well described as the *water-fever*. This is often to be regarded as an indication, that the baths are acting upon and influencing the system and the morbid condition, and that the use of them will be eventually beneficial ; the question for consideration being, perhaps, whether their use should be interrupted, or whether they should be used more sparingly, or whether the use of the warm baths should be substituted for that of the natural baths, either temporarily or otherwise. These more powerful effects of the baths are usually controlled by using them only every second, or every third day, or on two successive days with the interruption of the third day ; and by regulating the time of remaining in the bath, according to the strength, excitability, and other peculiarities of individual cases. Such restrictions have, moreover, an additional object. Whether the primary excitement from the use of the baths is considerable in its degree or otherwise, and whether the water-fever is evidenced strongly or otherwise, a course of these baths is almost always followed by some degree of general debility. This is first noticeable in the circulation: the pulse at the wrist, and the heart's

action become more feeble; but it is early marked by
languor and feebleness, and indisposition to make any
exertion, and despondency, diminished appetite, and disturbed or lethargic sleep.   The degree of this secondary
effect of the baths is usually inconsiderable; and it is
generally of short duration, when the baths are made use of
with the interruption of certain days, when the course is not
unwisely prolonged, and when the several immersions have not
been for too long a time.   But it is right to say, that when
such precautions are not used, these effects are sometimes
so great as to be of serious importance.   It would not be
to give a fair account of the medicinal effect of the Buxton
baths, nor to offer a caution which is often needful, as to
the use of the baths, if this were not thus stated in unqualified terms.   It often happens, that strong and otherwise
healthy individuals, suffering only from localised rheumatism
of some part of the trunk of the body, or of the limbs,
referable to exposure to cold and wet,—as in the instance
of miners, who frequently have to lie down at their work,
with one or both legs, and perhaps one hip, and perhaps
even one side of the body, covered with water, and this
during days and even weeks in succession,—are tempted,
from an anxiety to obtain relief, and to be in a condition to
return to their homes as soon as possible, to bathe more
frequently than is advised, or to remain longer in the bath
than is directed, with the consequence of a sudden and great
prostration of power, sometimes resulting in important and
serious disease.   Within a single week, such strong and

even athletic patients, without appreciable disturbance or deranged function of any internal viscus, with an ascertained healthy condition of the circulation, respiration, membranes, and faculties, have presented themselves in a state of much languor and exhaustion, evidenced by the condition of the heart's action, and every other least mistakeable indication ; and all this extreme effect been referable only to the use of the bath daily, and the having remained in the bath every time from five to fifteen minutes longer than had been desired.

Such effects as these are not experienced after the use of baths of ordinary water at any temperature, or repeated or continued to any extent; and the inquiry is naturally and at once suggested, as to the cause to which these effects are to be ascribed. The chemical constitution of these waters has therefore been a subject of speculation and inquiry, from the earliest records. It can be no subject of surprise, that every generation of men, seeing the great and marvellous healing powers of these waters, should have become dissatisfied with the investigations as to their chemical ingredients which had hitherto been made, feeling their utter inadequacy to explain so considerable an amount of medicinal effect. And thus, during the long space of nearly three hundred years, have these waters been the subject of anxious and painstaking investigation, to the successive generations of chemists and medical men ; leaving every succeeding race of investigators virtually as unable as before to explain satisfactorily and conclusively the effect of the waters, by reference to their ascertained composition. And yet the confidence expressed by the successive generations of medical observers, has been unvarying; and the kinds of disease for the relief of which they are found to be so useful, are the same as in the earliest times. The temperature, and the flow, and the clearness and brightness, and the freedom from smell or very marked taste, have been no less unvarying, than the effects on certain morbid conditions of the human system, and than the chemical constituents, whatever these may be.

After the imperfect investigation which the state of science in that age enabled him to make, Dr. Jones, in his curious book (published in the year 1572), is obliged to content himself with the conclusion, that the qualities of the water are due to the presence of "some excellent ore, rather than either brimstone, alum, bitumen, iron, copper, or any such like, for then it should in drinking be perceived by the taste. Albeit true it is, as affirmeth Galen, all such hot baths of such minerals have force of drying, but in these you have no such sense, but so fair, so pleasant, and so delectable, that it would seem to be a dulce bath made by art, rather than by nature; howbeit the effects declare brimstone to be therein. Sea water often strained through sand, becomes sweet, and so may these waters being strained through the earth, lose their mineral taste, but retain great virtue both manifest and hidden." Here may be remarked, even in those earlier days, when analytical chemistry could do so little, the full admission as to the powerful effect of the waters upon disease, that this effect must be ascribed to some medicinal constituent, and the conjecture that this constituent might be sulphur, or some similar agent, deprived by filtration, or some equally powerful means, of its taste and smell, but being left in other respects in an efficient conditon. Of the long catalogue of strangely named ailments, for which this water was then said to be curative, Dr. Jones places "Rheums" (rheumatism) at the head, as is done at the present time; and in the list, there are female weakness and irregularities, relaxed and irritable states of the mucous membranes, with their many and various morbid consequences: most clearly indicating, that the larger proportion of the invalids at that time resorting to the Buxton baths, were suffering from the same disordered conditions of system, as the greater number of those who make use of them at this time for the relief of their ailments.

Dr. Lister, the second in date of the ancient investigators of the Buxton water, any traces of whose works have been preserved, describes the medicinal effects of the baths as being stimulating; and states that, if too long continued, they produce wasting, feverishness, and debility; referring the effects to a small proportion of iron, which he affirms that he could taste, but could not otherwise detect; and stating, moreover, that the water contains a small proportion of common salt, and of calcareous earth.

Dr. Leigh, who wrote previously to 1671, testifies to the "surprising effects" which he had observed from the use of the baths, in cases of rheumatism. He says, " persons that could not go before without the help of crutches, came from thence to Manchester on foot without them, viz., sixteen miles." The distance, according to the modern measurement, is twenty-four miles; and it will be noticed, that ancient authors generally, in mentioning the distances from place to place in the district, state them as being two-thirds of the distances as estimated by modern measurement. Dr. Leigh ascribes the medicinal effect of the water to "marine salt, and the sal catharticum amarum, with the nitrum calcarium."

Dr. Short, in 1733, made a much more careful chemical examination of the Buxton tepid water, than any which had been made previously. He says, in his preface: "Many of the (mineral) waters in use have been so superficially examined, that it is impossible to draw any certain conclusion concerning their contents, or what they are or are not; and therefore they should be more thoroughly searched and sifted; Buxton, for instance, which though it has justly maintained its character these two thousand years, yet has there no pains been taken to discover its impregnating principles, except by Dr. John Jones (a Welshman, who lived some time at King's Meadow, near Derby), near two hundred years ago; and a transient visit made to it by Dr. Lister. Matlock, though much frequented of late years, yet the world are strangers to its contents, though some would have us believe that its virtues are exactly the same with those of Bristol (Clifton); but offer neither argument nor experiment to support their opinion." He says: "Since these waters continually bring up so large and

numerous bubbles with an impetuous force from the bowels
of the earth, then must their interstices be richly stored
with this fine air;" and he seems to have been inclined to
refer the medical action of the waters, either directly or
indirectly, to this air; but would not appear to have endeavoured to estimate the proportion of the air contained in
the waters, nor to ascertain its character. He ascribes the
principal effects of the water to "its warmth and mineral
spirit;" stating that he could not refer the medicinal effects
to the solid chemical constituents, which he computed to be
only 26 grains in the imperial gallon, of which he estimated
13 grains to be calcareous, and the remainder to consist of
marine salt and nitre in equal proportions. Dr. Short
testifies to the good effects which he had seen from the use
of the baths, in cases of gout and rheumatism; and then says,
that he would refer the medicinal action of the waters to
"a subtle mineral principle or spirit, wrapt up in the air
(contained in them)." He mentions the favourable effects
from these baths, in cases of contraction and stiffness of the
limbs, the consequence of "rheumatic and arthritic pains."
He concludes that the water is highly impregnated with a
mineral steam, vapour, or spirit, containing a most subtle
and impalpable sulphur; herein following the idea suggested
by Dr. Jones so long before; but not seeming to be con-
scious that he had borrowed the theory from any preceding
writer, although he had quoted largely from Dr. Jones's
work. Dr. Short mentions the effects which he had
witnessed from the use of the baths, in relieving uterine

obstructions, in removing periosteal thickenings, in removing
the effects of old sprains, in affording relief to certain
disordered conditions of the kidneys and bladder; and he
advises that both the baths and the drinking of the waters
should be used with discrimination; by no means always
or necessarily drinking the waters and using the bath at the
same time, or in all cases; but drinking the waters in some
cases, bathing in others, and in some cases using the waters
in both ways. Dr. Short says also, "let me add once for all,
that as this water is of such a nature as I have mentioned,
so it is not to be trifled with, for if it be unnecessarily used,
it will certainly do harm;" and he judiciously adds, that the
use of the waters is not advisable in inflammatory cases,
nor "in consumptions attended with a rapid motion of the
blood, and weak pulmonary vessels."

Dr. Hunter published a "Treatise on the Nature and
Virtues of the Waters of Buxton," in 1765. The results of
his analysis are nearly the same as those of Dr. Short. His
estimate of the proportion of "calcareous earth" is somewhat larger than that of Dr. Short, and of the proportion
of "sea salt" and "native alkali," is somewhat smaller;
but his total results, as to the amount of the saline constituents, and their nature, are nearly the same.

Dr. Percival made "Experiments and Observations on the
Buxton Waters," which were published in the sixty-second
volume of the "Philosophical Transactions." The estimate
of the total saline constituents deducible from these experiments, is nearly the same as that of Dr. Short; and they

are referred to the headings—sea salt, calcareous earth, and alkali.

Dr. Higgins published an analysis of the Buxton waters in 1782, and, so far as the solid constituents of the water, the analysis is singularly successful. With the needful correction, to make the result correspond with the imperial gallon of the water, it would be as follows :—

	Grains.
Sea salt	4·6
Calcareous earth, combined with acidulous gas .	15·1
Sulphate of lime	2·0
Chloride of magnesium	1·6
Iron, combined with acidulous gas . . .	0·6
Saline constituents in an imperial gallon .	23·9

This appears to have been the only instance, previously to Dr. Lyon Playfair's analysis, in which any trace of iron was detected in these waters.

In 1784, Dr. Pearson published his great and excellent work, entitled " Observations and Experiments for investigating the Chymical History of the Tepid Springs of Buxton." It is to this analysis, that the discovery of the nature of the gaseous impregnation of the waters is to be referred. The following paragraph is a summary of the more important of Dr. Pearson's observations and discoveries, which may be supposed to be still of general interest, or to bear upon the character and properties of the water in the present day.

" The water is of crystalline transparency, and is colour-

less. When a large bulk of it is viewed together, as that
contained in the baths, where it is four or five feet deep, it
is colourless, and objects may be seen through it. This
crystalline fluid exhibits bright bubbles, of the size of the
smallest pin's head, adhering to the sides of any vessel
containing it, or whatever is immersed in it. The baths
contain these bubbles in every part of them, especially upon
a little agitation. Moreover, streams or clusters of these
bubbles, of various sizes, from the magnitude of the smallest
pin's head to the bulk of a cherry, or even sometimes of a
billiard ball, every now and then break out from the floorings
of the baths, and dart perpendicularly upward, through the
whole thickness of the water. In a portion of the water
that has ceased to manifest bubbles in a temperate heat,
by exposing it to a greater degree of heat, they will again
appear. There is no smell from this fluid, nor will it
become fetid by standing, as some have asserted. It is
perfectly insipid; in particular, it has not the slightest
acidulous taste. The temperature is 81¼ to 82 degrees,
according to Fahrenheit's scale. There is every reason to
believe that this water has been of precisely the same temperature for many hundred, perhaps many thousand years.
It has certainly been of the same heat, since this property
was first determined by the use of specifically graduated
thermometers, more than thirty years ago (1750-53).
When the bath is agitated, as by the plunging of the bathers,
the transparency of the water is changed to that of turbid-
ness; but as soon as the commotion subsides, it becomes

instantly clear as before. This turbid appearance has been ascribed to impurities or to sedimentary matters deposited on the pavement of the bath, and stirred up and mixed with the water; but it certainly is not occasioned by this circumstance, because it may be produced at all times, even immediately after the bath has been thoroughly cleansed and re-filled, and when there is no sediment either observable, or by any possibility present. Moreover, when glass vessels were filled with this turbid water, it appeared perfectly clear, nor did it deposit any sediment on standing. (This muddy appearance is no doubt referable to the large quantity of gas, that is mixed with or suspended in the water). The medicinal qualities of this water chiefly depend upon a permanent vapour. This permanent vapour (gas) is inodorous, is not acidulous, occurs in exceedingly minute bubbles, which are diffused throughout the whole bulk of the water, and are not by any means merely adherent to the sides of the vessel containing it. This vapour (gas) is elastic, yielding to pressure, and recovering its former volume when the pressure is removed. It continues as a vapour at all temperatures, and is colourless. It cannot support combustion. The gleam of a taper introduced into it was constantly extinguished. Animal life is supported by it or maintained in it, during a much shorter time than if allowed to respire an equal amount of atmospheric air. Many kinds of water contain more atmospheric air than this water, and many kinds of water contain more carbonic acid gas; but none appear to contain the same amount as this water, of this peculiar, elastic, and aëriform principle. We are instructed, as the results of many experiments, that this water does not contain any impregnation which is evidenced to the senses, except the permanent vapour, which is not carbonic acid gas, nor any vapour which is odorous,—that the heat of the water much exceeds the temperature of the ordinary springs of the district; the temperature of such springs being usually about 48 or 50 degrees;—and that the water contains acid of vitriol (sulphuric acid) and marine acid (muriatic acid), combined with lime and alkali, and carbonic acid combined with lime, in addition to its impregnation with the permanent vapour."

Dr. Pearson estimated the gaseous impregnation to be only $\frac{1}{14}$th part of the bulk of the water, at ordinary temperature, and under ordinary pressure; and the following was the result of his analysis of the solid ingredients—

	In the Imperial Gallon. Grains.
Chloride of sodium	2·333
Sulphate of lime	3·333
Carbonate of lime	14·000
Total solid ingredients per imperial gallon	19·666

or 19½ grains.

In the year 1819, or thirty-five years after the date of Dr. Pearson's analysis, Sir Charles Scudamore and Mr. Garden jointly examined the Buxton tepid waters, with the appliances and greater accuracy of the more advanced state of science.

According to the careful and excellent analysis performed by these gentlemen, the imperial gallon of the waters was estimated to contain—

	Grains.
Chloride of magnesium	·773
Chloride of sodium	3·200
Sulphate of lime	·800
Carbonate of lime	13·866
Extractive matter	·666
Loss	·693
	19·998

or 20 grains of solid or saline matter.

The examination of the gaseous impregnation of the water, served to confirm Dr. Pearson's discovery as to the nature of the gas. In the words of Sir Charles Scudamore, "Dr. Pearson found that the proportion of carbonic acid, in the Buxton water, did not exceed the half of what is found in many common springs. He had the merit of discovering the separate existence of azote in this water, a principle which had never been detected by any preceding chemist in any water. In the imperfect state of chemistry, thirty-six years ago (1783-1819), the nature of azote was unknown, and he described it, ' as being a permanent vapour, composed probably of air and phlogiston.' The present analysis gave about one-fifth more of azote in a gallon, than appears from Dr. Pearson's conclusions."*

According to Sir Charles Scudamore's and Mr. Garden's

---

* A Treatise on Mineral Waters. 2nd Edition. London, 1833.

G

analysis, the imperial gallon of the waters appeared to contain of gaseous impregnation—

	Cubic Inches.
Carbonic acid	2·00
Nitrogen	6·18
Total	8·18

The proportion of nitrogen was supposed to be rather more than three times that of the carbonic acid contained in the waters.

In 1852, the water was analysed by Dr. Lyon Playfair, with the subjoined result.

Analytical Report on the Water of the Thermal Springs of Buxton, by Dr. Lyon Playfair, C.B.,F.R.S.

"MUSEUM OF PRACTICAL GEOLOGY AND GOVERNMENT SCHOOL OF MINES.

"*London, July 24th,* 1852.

"To SIDNEY SMITHERS, ESQ.

"Sir,—In consequence of a request made by you, on behalf of his Grace the Duke of Devonshire, I visited Buxton on the 8th and 9th of April, for the purpose of collecting the water of the thermal spring for analysis.

"The water was collected partly in glass-stoppered bottles, and partly in earthenware jars. The gas, as it issued from the crevices of the rock and bubbled through the water, was caught by an inverted funnel, and collected in glass bottles filled with the thermal water itself. These bottles were then sealed on the spot; and the evidence derived from the gas contained in them, shows that the precautions used for preventing the access of air were quite successful.

"It is not necessary for me to describe the physical conditions under which the thermal springs appear at Buxton. It may be sufficient to state, that they issue from fissures in the limestone, and are accompanied by frequent, but intermittent bursts of gas, which escapes partly as large bubbles, and partly in innumerable small bubbles, giving to water freshly collected in glass vessels, all the appearance of soda water.

" The water is clear, sparkling, inodorous, and when cool, is almost tasteless. Its temperature is 82° Fahrenheit, and its specific gravity 1·0003.

"Two points had specially to be attended to in the analysis of the waters,—firstly, to ascertain the nature and quantities of the ingredients in solution, and, secondly, the character and composition of the gas accompanying them.

In order to be sure that every ingredient came under my observation, I caused 100 gallons of the water to be evaporated down to about half a gallon, and examined the deposit and residual solution for bodies which might be present in such small quantity as to escape detection in the unconcentrated water. The precaution was found to have been necessary, for, in addition to the ordinary constituents of the waters, two more rarely occurring bodies—viz. fluorine and phosphoric acid—were found to be present, although only in minute quantity. The amount of fluorine was, however, sufficient to etch glass when applied with proper precautions. Neither iodine nor bromine could be detected.

" The following analysis gives the amount and nature of the solid ingredients in one imperial gallon of the water at 60° :—

	Grains.
Silica	0·666
Oxide of iron and alumina	0·240
Carbonate of lime	7·773
Sulphate of lime	2·323
Carbonate of magnesia	4·543
Chloride of magnesium	0·114
Chloride of sodium	2·420
Chloride of potassium	2·500
Fluorine (as fluoride of calcium)	trace
Phosphoric acid (as phosphate of lime)	trace
	20·579

" On examining the water, there were found present carbonic acid and nitrogen, in addition to the solid ingredients. It was important to estimate the amount of the former in an exact manner. Some of the water was received from the spring into a glass-stoppered bottle, and the stopper was immediately inserted and secured. One gallon of the water was found to contain altogether 13·164 grains of carbonic acid; but of this quantity, 5·762 grains were due to the carbonates of lime and magnesia, and therefore only 7·402 grains could in any sense be considered as free. Again, the carbonates of lime and magnesia are present as bicarbonates, or as carbonates dissolved in carbonic

acid, and 5·762 grains of carbonic acid would require
to be added for this purpose.   Hence of the 7·402 grains, or
15·66 cubic inches of gaseous carbonic acid in the water,
only 1·640 grain, or 3·47 cubic inches, can be considered
as wholly free and uncombined.

"The nitrogen in the water could only be present in
solution, and not in combination; and as there is no very
accurate method for ascertaining the precise quantity of this
gas in the water at any given temperature, it was considered
chiefly important to ascertain accurately the composition of
the escaping gas, as this would indicate that of the gas held
in solution.   The following are the analyses of two portions
of the gas collected as formerly described, the analyses
being given *according to volume*.

	I.	II.	Mean.
Carbonic acid . .	1·169	1·164	1·167
Nitrogen . . .	98·831	98·836	98·833
Oxygen . . .	trace	trace	trace
	100·000	100·000	100·000

"The gas, therefore, consists entirely of carbonic acid and
nitrogen; for the oxygen, which did not amount to one-tenth
per cent., may be viewed as quite accidental, arising probably from the corks used to close the bottles.

"Judging from the analysis and proportion of the gases,
it is assumed that *at the moment of issue*, the water is
charged with 206 cubic inches of nitrogen, and 15·66 cubic
inches of carbonic acid.   This assumption is founded upon
the proportional relation of the two gases.   The proportion

of carbonic acid in the water being determined, and the proportion of carbonic acid to that of nitrogen contained in the water being 1·2 to 98·8, the amount of nitrogen contained in the water at the moment of issue may fairly be assumed to be 206 cubic inches per gallon.

" Before remarking further on the above analysis, it may be useful to refer to that by Scudamore. The analysis given by him was upon the wine gallon, which is one-fourth less than the imperial gallon. Correcting for this difference Scudamore found twenty grains of solid matter in a gallon— a result not materially different from that detailed above. The solid ingredients do indeed differ to some extent in the two analyses; but it must be recollected that analytical chemistry is now in a much more advanced state; and instead of being surprised at the differences, we are rather inclined to admire the precision with which the points had been made out.

" From a consideration of the previous analysis, I am inclined to ascribe the medicinal effects of the water almost entirely to its gaseous constituents. The water, deprived of its gases, has the composition of an ordinary spring water, with the exception of the fluorine and phosphoric acid, both of which are present in mere traces; and it is therefore difficult to conceive that they can have any medicinal effect when the water is used for baths. The gases are, however, nearly of the same composition as those of the thermal spring at Bath, and there is no reason to doubt that dissolved carbonic acid and nitrogen may exert important physiological effects. At all events, the singular chemical character of the Buxton tepid water must be ascribed to its gaseous and not to its solid ingredients.

"Sir,

"I have the honour to be

"Your obedient and faithful servant,

"LYON PLAYFAIR, F.R.S."

.. The different analyses of these waters, which have been made at different and distant periods down to the present time, have thus been set forth, in order to indicate the difficulties which have at all times attended the rationale of their effects in disease; and to show how early the opinion came to be entertained, that these effects might be ascribed, in a principal or important degree, to the character and quality of the gas, vapour, or halitus, which is contained in the water. It is admitted to be still difficult to determine the precise nature or extent of the effects of uncombined nitrogen, when introduced into the human system, whether by absorption through the skin, or through the mucous membrane of the stomach. It may even be true, that the whole of the medicinal effect of the water is not due to the nitrogen which it contains. It may be referable to some constituent, which even the greatly advanced state of modern chemistry has not been able to detect. But it is just to indicate, that the medicinal action of nitrogen may not be unequal to produce great medicinal effects, when so exhibited as to be absorbed into the system

with great readiness, and in large amount. The effect of
nitrogen throughout the economy of the earth, is now known
to be very great and all-important. The agent which, only
a few years ago, was considered to be simply a great diluent
of the oxygen in the atmosphere, and to have only the effect
of lessening the action of this great stimulating and oxydi-
sing principle, nitrogen is now ascertained to be an
important component of many animal substances, and an
indispensable element in the nutriment of animal life.
There is no single particular, in which the laborious and
successful investigations of modern chemists, and the application of the results to physiology and pathology, have been
so influential and important, as in the developement and
elucidation of the importance of nitrogen, in its multiplied
combined relations to the phenomena of life. The high
authority of Baron Liebig may be quoted in support of this
statement, in reference to its different and important
bearings, by adducing a succession of sentences from
his works. "All parts of the animal body which have
a decided shape, which form parts of organs, contain
nitrogen. No part of an organ which possesses motion
and life is destitute of nitrogen."—"All kinds of food
fit for the production either of blood, or of cellular tissue,
membranes, skin, hair, muscular fibre, &c., must contain
a certain amount of nitrogen."—"Water and common
fat are those ingredients of the body which are destitute
of nitrogen. Both are amorphous or unorganised, and only
so far take part in the vital process as that their

presence is required for the due performance of the vital functions."—"All such parts of vegetables as can afford nutriment to animals, contain certain constituents which are rich in nitrogen; and the most ordinary experience proves, that animals require for their support and nutrition less of those parts of plants, in proportion as they abound in the nitrogenised constituents."—"The chief ingredients of the blood contain 17 per cent. of nitrogen, and no part of an organ contains less than 17 per cent. of nitrogen."—"All experience proves, that there is, in the organism, only one source of mechanical power; and this source is the conversion of living parts into lifeless, amorphous compounds." —"No part of the body, having an organised or peculiar form, contains, for 8 equivalents of carbon, less than 1 of nitrogen."—"Out of the newly-formed blood, those parts of organs which have undergone metamorphoses are reproduced. The carbon and nitrogen of the food thus become constituent parts of organs. Exactly as much sulphur, carbon, hydrogen, and nitrogen is supplied to the organs by the blood—that is, ultimately, by the food—as they have lost by the transformations attending the exercise of their functions."—"The flesh and blood consumed as food, ultimately yield the greater part of their carbon for the support of the respiratory process, while the nitrogen appears as urea or uric acid, the sulphur as sulphuric acid. But previously to these final changes, the dead flesh and blood become living flesh and blood; and it is, strictly speaking, the combustible elements of the compounds formed in the

metamorphoses of living tissues, which, with some other
substances, to be more particularly mentioned hereafter,
serve for the production of animal heat."

  These quotations may serve to illustrate and justify the
degree of importance ascribed to nitrogen, in the phenomena
of life,—in the nutrition and expenditure of the animal
economy. Every movement of the animal machine involves
the expenditure of some portion of the existing and living
tissue; and every such expenditure involves the consumption
of a given proportion of nitrogen, and demands its restoration in the form of aliment, in the composition of which
nitrogen is an essential element. The nitrogen, to be thus
useful, must be combined with other elementary substances,
and combined in certain proportions; but such compounds
do not exist without nitrogen; and this element is essential
to organic structure, to animal function and movement, and
to nutrition.

  Moreover, to return to the words of Baron Liebig,
"Medicinal or remedial agents may be divided into
two classes, the nitrogenised and the non-nitrogenised.
The nitrogenised vegetable principles, whose composition
differs from that of the proper nitrogenised elements of
nutrition, also produced by a vegetable organism, are distinguished, beyond all others, for their powerful action
on the animal economy. The effects of these substances
are singularly varied; from the mildest form of the action of
aloes, to the most terrible poison, strychnia, we observe an
endless variety of different actions. With the exception of

three, all these substances produce diseased conditions in the healthy organism, and are poisonous in certain doses. Most of them are, chemically speaking, basic or alkaline. No remedy, devoid of nitrogen, possesses a poisonous action in a similar dose. This consideration, or comparative view, has led to a more accurate investigation of the composition of picrotoxine, the poisonous principle of cocculus indicus; and Mr. Francis has discovered the existence of nitrogen in it, hitherto overlooked, and has likewise determined its amount."

In these instances likewise, the nitrogen is in combination; and it is in virtue of the proportions of such combination, that the resulting compounds are thus powerful in their effects on the animal economy; but the nitrogen is essential to the result, and it is not a mere diluent.

Once again,—"*Disease* occurs when the sum of vital force, which tends to neutralise all causes of disturbance (in other words, when the resistance offered by the vital force), is weaker than the acting cause of disturbance."— "In medicine, every abnormal condition of supply or of waste, in all parts or in a single part of the body, is called disease."—BARON LIEBIG.

Such illustrations, cited from such authority, manifest the great importance of nitrogen in the economy of life, and in the production and the cure of disease.

" The Edinburgh Medical and Surgical Journal," No. 193, October, 1852, in an elaborate and able analytical review of my " Letter to Dr. Lyon Playfair," contains the following passage, and also the well-selected quotation from Dr. Sutro's excellent "Lectures on the German Mineral Waters."

" The thermal spring of Wildbad in the Black Forest contains, with a minute amount of saline matters, a large amount of nitrogen, 80 per cent.; and to the presence of this gas, the German physicians and Dr. Sutro ascribe the curative effects which the use of the Wildbad water exerts upon chronic rheumatism, rheumatic gout, and stiffness and nodosity of the joints. So also the waters of Pfeffers in the Canton of St. Gallen in Switzerland, and that of Gastein in the Mountains of Salzburg, contain, the former a small proportion of nitrogen, the latter a good deal more (2·02 in 100 parts of water). It seems therefore quite natural to ascribe to the presence of this gas very notable effects upon the human organism; though in what exact manner these effects are produced, it is not so easy to understand and explain.

".The opinion of Dr. Sutro is given in the following passage :—' Without oxygen we should suffocate, without nitrogen we should starve. I should not go so far as to attribute a nourishing property to the nitrogen introduced into the absorbent vessels with the highly-diluted water. But when it is admitted, on all hands, that our tissues constantly discharge wasted particles in proportion to the regular additions provided by .the arterial supplies; and when we know a great part of this waste to issue from our cutaneous pores in a gaseous form, would it not be reasonable to attribute some restoratory function to the contact and combination of the gas with organic particles? We know that, in old age, earthy or inorganic formations prevail in the reproductive sphere. Limbs become more rigid, the joints less pliable, secretions retarded, excretions diminished, vital elasticity and resisting power impaired. Substances ordinarily carried rapidly along the vascular canals in a dissolved state, are now precipitated out of the slowly moving mass, and deposited in spaces where they further impede voluntary movement.'

"If we see the use of a mineral water, causing distinct retrogression of these anti-vital phenomena; if we perceive gouty concretions to proceed towards absorption; if we observe contracted limbs gradually to relax again, and to try feeble efforts of long-forgotten exercise; if we find cutaneous harshness and rigidity to diminish, and to give way to a former softness; if we behold a resuscitated desire for muscular exertion and for mental work in a prostrate individual, and we know the spa, the originator of these changes, to possess a great quantity of nitrogen, is it not legitimate to attribute to this gas part of the efficacy?"

It is evident, then, that the result of Dr. Lyon Playfair's analysis justifies a much enlarged expectation as to the medicinal value of the Buxton tepid waters, based upon what is known as to their chemical constitution. The great and unlooked-for discovery, that these waters may be fairly assumed to contain 206 cubic inches of nitrogen per gallon, at the moment of their issue, leads to unavoidable inferences as to their medicinal value and importance. And as the chemistry of healthy structure and function, and of diseased conditions, attains a more advanced and influential and confirmed position, it is probable that so great an elementary principle as nitrogen, poured forth in such vast quantity, and with such ceaseless rapidity, and in a form so readily available for internal or external use as the Buxton tepid waters, may acquire a greater and greater value; adding to the prestige and the fame due to their ascertained effect on disease, the confirmation and precision of sound theoretical data; and enabling their use to be extended to any and all the various forms of disordered action, in which the direct supply and influence of nitrogen to the fluids, or the tissues, or affected organs of the body, might be ascertained to be a direct mode of antagonising disease. It seems to be probable, that the great effect of these waters upon some diseases, which has been so long known, and so largely appreciated, may be thus accounted for; and it may probably be hoped, that, in its turn, the effect of introducing so much free nitrogen into the system may help to explain the nature of the diseases on which these waters act so energetically. So long as 6·18 cubic inches of nitrogen per imperial gallon, were supposed to be the whole amount of this important element contained in the waters, it appeared to be difficult or impossible to ascribe to it the great medicinal effect produced by the use of the Buxton baths, or by drinking these waters. The recent analysis has placed this question in a very different position.

As to the solid constituents of the waters, it is only
indirectly that the result of the recent analysis can be said
to be of much importance.   It is indeed needful, and
only the just due of a mineral water, to which the long-
continued and large resort of sufferers from rheumatism,
gout, &c., attach much importance, that the more advanced
state of chemistry should be brought to bear upon it
from time to time, in order to determine whether, or to
what extent, additional discoveries as to the substances
which enter into its composition, may bear out, explain,
or extend its usefulness and applicability in different
diseases.   The result is that silica, oxide of iron, alumina,
fluorine, and phosphoric acid have been for the first time
ascertained to be among the substances dissolved in the
waters.   The proportion of these constituents is indeed
small.   But since the presence of these ingredients in the
waters had not been detected, even by the analysis which was
made with so much care and skill by Sir Charles Scudamore
and Mr. Garden, at a time comparatively so recent, and
with all the means and appliances which chemistry possessed, it seems to be an unavoidable inference, that as this
rapidly advancing science attains greater and greater
perfection in its processes and teachings, it may help us
to explain more and more satisfactorily the means, by virtue
of which these waters act so usefully in the relief of disease.
Such explanation may prove to be derivable exclusively,
from the effect that may be referred to the direct introduction of so much free nitrogen into the animal economy,

by the use of these waters, whether externally or internally ;
or it may be assigned, in part, to the introduction, in a
peculiarly available state of combination and dilution, of the
solid ingredients already ascertained to be contained in the
waters; or it may come to be partially referred to a constituent or constituents which have not hitherto been
detected in it. Looking at the great advance which has
been made in the science of chemistry, in the minute
accuracy of its manipulations and results, in the closeness
of its reasonings, the breadth of its deductions, the value
and bearing of its inferences, and its extensive and much
extended influences on all collateral branches of science,
both in medicine and the arts, it is impossible to doubt that
more certainty may be obtained as to the *modus operandi*
of these waters in disease, than we now possess. And yet, the
facts already ascertained are so important and conclusive, in
regard to the solid and gaseous constituents of the waters, as
to warrant a full *à priori* confidence in its medicinal
character. By Dr. Pearson's analysis, carbonic and sulphuric acids, chlorine, sodium, lime, and free nitrogen, were
ascertained to be contained in them; this analysis was confirmed, and the presence of magnesium detected by Sir Charles
Scudamore and Mr. Garden; and by Dr. Lyon Playfair's
analysis, the presence of silica, iron, alumina, potassium,
fluorine, and phosphorus was ascertained. So large an
amount of additional information commands additional confidence, and serves to confirm and establish the character
and value of the waters, independently of theories, and in

aid of the immemorial experience of their medicinal efficacy.

Instead of 2 cubic inches of carbonic acid in the imperial gallon of the waters, which was the result in the immediately preceding analysis, the recent analysis shows that there are 3·47 cubic inches of this gas in the gallon, after having deducted for every form in which the remaining 12·19 cubic inches obtained from it may be supposed to be held in combination. But free carbonic acid, even when contained in very large proportion in mineral waters, is not found to have much medicinal effect. Carbonic acid is chiefly valuable in mineral waters as a solvent for more powerful ingredients, and as a means by which the more rapid absorption of the waters, either through the skin or through the stomach, is secured. The effect of carbonic acid in these less direct ways, is by no means unimportant; and, so far, the larger proportion of this gas, now ascertained to be contained in the waters, is worthy of notice; but, after all, the amount of carbonic acid contained in the waters is small, when compared with that which is contained in many mineral waters.

That no less than 206 cubic inches of free nitrogen may be fairly assumed to be present in the imperial gallon, is a much more extraordinary and interesting fact; and no less extraordinary and interesting when viewed in connection with the great flow of the waters, than when compared with the results of the previous analysis. Supposing, as may be probable, that the whole flow of the waters is 300 gallons per

minute, the amount of free nitrogen discharged along with
them is 61,800 cubicinches, or nearly 36 cubic feet per minute;
and if, at a moderate estimate, 150 gallons of the waters pass
every minute through the baths and the wells, then 30,900
cubic inches per minute of this important elementary
principle, with whatever medicinal action nitrogen may
subserve, are offered in an available form for either external
or internal use.   The knowledge of this great fact, resulting
from Dr. Playfair's analysis, is partly due to the improved
methods now possessed for ascertaining the character and
proportions of gases; and it should be ascribed partly to
the great care which was taken, to secure the whole of the
gaseous contents in the waters submitted to analysis.   And
if the free application of uncombined nitrogen to the surface
of the body, or to the lining membrane of the stomach, is
capable of influencing the human system in any degree, the
medicinal effect of these waters cannot but be held to be so
far fully explained.   The proportion is so much greater than
has hitherto been claimed as being contained in any other
mineral waters whatever, that the belief in their medicinal
character cannot but be so far strengthened in an important
degree.   These waters have been more frequently compared
and likened to the important mineral waters of Wildbad, in
Wurtemburg, than any other of the great continental thermal waters, inasmuch as they seem to be used with success
in many of the same disordered states of system, and inasmuch as the saline constituents of the two waters are in
many respects similar, and as the Wildbad waters contain

a considerable proportion of nitrogen. But whereas 100 parts of the gas obtained from the Buxton waters contain little more than one part of carbonic acid, and only a trace of oxygen, nearly 99 parts consisting of free nitrogen, 100 parts of the gas attained from the Wildbad waters contain 12½ parts of carbonic acid, and 8¼ parts of oxygen, 79¼ parts only being nitrogen; and of this smaller proportion of nitrogen, 36 parts, or nearly one-half, must be deducted as corresponding with the 8¼ parts of oxygen, and representing so much atmospheric air. The absence of oxygen in the gas obtained from the Buxton waters enhances the estimate of the nitrogen obtained from it, inasmuch as the nitrogen is thus left free, and available for any purpose, medicinal or otherwise, that so much free nitrogen may be supposed to serve. And, according to the respective analyses, the Buxton waters are assumed to contain more than half the proportion of free nitrogen that the well-known Seltzer waters contain of carbonic acid—a statement which conveys fully and clearly, a notion of the very large proportion of nitrogen which is contained in the waters of the Buxton tepid springs. It may seem to be invidious to compare these waters with other warm mineral waters; but it is surely just to set forth its claims, not only to a distinguished, but, as it would thus far appear, to a first position, in regard to the amount of its impregnation with free nitrogen gas, and in regard to whatever medicinal value may be believed to attach to it on this account.

It seems to be sufficient, and to be as much as consists

with the present state of information, to have learned that
these waters contain these saline and gaseous constituents,—
to ascertain what effects the use of the waters, as baths and
internally, produces on the human system, in health, and in
different disordered or diseased conditions,—and to assume
that the effects must be referable to what has been ascertained as to the constituents of the waters. It would have
been as impossible to infer *à priori*, that a certain proportional combination of three or four elementary substances
would produce an alimentary substance, a certain other
combined proportion of the same elementary substances
would produce a substance having valuable medicinal properties, a third proportional combination of the same substances produce a virulently poisonous compound, and a
fourth compound of the same ingredients produce a substance that would be neither alimentary, nor medicinal, nor
poisonous, but a substance insoluble in the gastric secretions,
and altogether inert when received into the human stomach.
And yet the chemistry of organic substances furnishes
many instances of this remarkable character, which the
present amount of our information leaves unexplained.
The same component elements, in different proportions, form
the most powerful of the vegetable tonics (quinine), the
most active of the vegetable narcotics (morphia), the most
powerful of the vegetable poisons (picrotoxine), and the most
valuable of alimentary restoratives (animal and vegetable
fibrine). The effects of these compounds are no less certain,
and the grounds for their use or avoidance, and the doses

and circumstances of their use, are no less trustworthy and defined, because the reason of such difference in property cannot be ascertained. The admission, that the degree of effect or the kind of effect on the system of the Buxton tepid waters, could not be predicated from the nature of their chemical constituents, is no invidious or singular admission of limited knowledge; nor can it affect the trust which science attaches to experience, when the peculiar character of the tepid mineral waters is thus established. An important amount of medicinal value may be claimed for them, on the exclusive ground of their chemical constitution.

It has been advanced, as a mode of explaining the medicinal action of the Buxton tepid waters upon the animal economy, that the absorption of the nitrogen with which they are so largely charged, leads to the formation of so much ammonia, by involving the decomposition of a due proportion of water to furnish the required amount of hydrogen; and that the ammonia thus formed, and brought to bear immediately upon the blood and tissues, is the essentially curative principle of these mineral waters. There is no foundation for this hypothesis; the supposed conversion of the nitrogen into ammonia is entirely conjectural, and extremely unlikely; and even if it were otherwise, the action of ammonia would be inadequate to explain that of the Buxton waters. These waters are more stimulating and more alterative in their effects, than could be accounted for in this way. This ammoniacal hypothesis is attempted

to be supported by a statement, that the diseases for which
the action of the Buxton waters is known to be remedial, are
marked by a deficiency of ammonia in the secretions.
Animal chemistry, however, demonstrates the incorrectness
of this assertion. Even the urine of healthy persons does
not contain so much ammonia as serves to neutralise the
acids which it contains, and urine ought always to show a
slight excess of acidity; and in almost all the diseases of
excitement, or of inflammatory character, the urine is likewise, in at least an equal degree, characterised by pre-
dominant acidity. But the ammonia which characterises
urine is, for the most part, formed by the putrescence of
the urea, and other highly animalised matters contained in
it, long after it has been discharged from the system. If
the remarkable similarity in composition of carbonate of
ammonia and urea be considered; and the fact, that a considerable proportion of the excess of uric acid in gouty and
rheumatic conditions, would seem to be obtained at the
expense of the urea, and to be the consequence of an imperfect decarbonisation of the blood during the process of
respiration, as ably urged by Dr. Gairdner in his excellent
treatise on Gout; the utter fallacy of this ammoniacal view,
as to the action of the Buxton waters, either as regards the
effects of ammonia, or the condition of disease, is fully
demonstrated. Medical men need not be told that ammonia
is equally inadequate for the relief of gout or the cure of
rheumatism, in whatever form it may be made use of:
complaints in which the efficacy of the Buxton waters is so

signally evidenced; and it is important that public attention should not be diverted from the fact of this efficacy, and from what is ascertained as to the chemical character and peculiarities of these waters, by speculations which are untenable.

CHAPTER V.

———◆———

THE whole of the town of Buxton, as has been said, lies
in a valley, and is surrounded by hills of greater elevation
than its own level. This applies more particularly to
Lower Buxton, which is immediately protected on the
south by St. Anne's cliff—now more commonly called the
Terrace-walks, and on the north by the rising grounds on
which the new church and the stables are placed. This part
of the town is well sheltered in all directions. It is more
immediately protected by plantations on the west; and on
the east by the higher grounds of Fairfield, and the rocks
which bound the valley through which the road to Bakewell
passes, close to the town. Upper Buxton is much less
sheltered; the higher grounds are situated at greater
distances, and its position is by so much one of greater
exposure. There is a difference of elevation between the
carriage road in front of the Crescent and the centre of the
market-place, amounting to seventy-six feet, nine inches;

and the elevation of Upper Buxton may therefore be said in round numbers to be 1100 feet, the elevation of the new church being 1029½ feet. In regard to the degree of shelter afforded to Upper Buxton, there is, however, within less than a mile, on the south, a range of ground which is three hundred and fifty feet higher; and at nearly the same distance, on the east and on the north, are grounds of as great or greater elevation. On the west, the two miles in length of the Buxton valley intervene between Upper Buxton and the higher grounds in that direction.

These several elevations, and the various elevations of the different more important positions throughout this district, have been obtained either from the excellent surveys published under the authority of her Majesty's Board of Ordnance, or from private surveys which have been kindly made in reference to this work, and the perfect accuracy of which may be implicitly relied upon. But an approach to relative accuracy may be obtained in a most interesting and ready manner, in regard to any locality, whether upland or valley, by the use of the very ingenious instrument—the Aneroid barometer. Barometers are used to indicate the pressure of the air; and therefore they may be had recourse to not only as weather-glasses, but inasmuch as they fall when higher ground is ascended, and rise when lower ground is descended,—the weight of the superincumbent atmosphere, by so much diminishing in the one case, and increasing in the other,—they act usefully in obtaining the relative elevations of different places above the level of the sea. The

H

Aneroid barometer is sufficiently portable to be conveniently made use of for this purpose. In using this instrument, it is only necessary to have obtained the elevation of any given object in a district; as, for instance, that of the New Church at Buxton. This is to be taken as the standard of the comparative observations, and the index of the Aneroid barometer is to be accurately read and noted on any given day, when the relative elevation of any other part of the district is wished to be ascertained. Every inch on the index of the instrument is divided into forty spaces, and every one of these spaces may be considered, with a sufficient approach to accuracy to satisfy most observers, to signify twenty-one feet. If any of the neighbouring eminences be then ascended, the index of the barometer will be found to fall more and more, as the higher and higher ground is attained; and by multiplying the number of spaces thus indicated by twenty-one, a sufficiently near approximation may be made to the relative elevation of any part of the district. Thus, for instance, it may be learned that there is a range of the index of nine and a half points between the level of the New Church and that part of the road to Manchester which is about three-quarters of a mile distant from the church, a little beyond the Royal Oak inn. Multiplied by twenty-one, a higher elevation is shown of 200 feet; or, if added to 1029½ feet, the elevation of the church above the level of the sea, the elevation of this part of the road is shown to be about 1230 feet. Again, between the level of the church and that of the highest part of the same road, called the top of the Long Hill, the index shows a fall of eighteen points, which when multiplied by twenty-one, gives a higher elevation of 378 feet, or a total elevation above the sea level of 1408 feet. In a district which presents so many different elevations of country and places, this instrument supplies an interesting and valuable resource to the tourist and the enquirer. It should be remembered, however, that such an estimate, although sufficiently near for most purposes, is only an approximation to the truth: the attainment of absolute accuracy by means of barometrical observations, requires some deductions for variations of temperature, and other influencing circumstances, and necessitates a somewhat intricate process of mathematical calculation.

The mineral waters, baths, wells, and their appurtenances, are situated in Lower Buxton, and contiguous to the Crescent. These are contained in two wings, at the east and west ends of the Crescent.

"The diameter of the inner circle on which the Crescent is built is about two hundred and forty feet, that of the outer one three hundred, and the breadth of each wing is about fifty-seven feet, making the length of the whole building nearly three hundred and sixty feet. The upper stories in the front are supported by an arcade, within which is a paved walk, about seven feet wide, where the company may take air and exercise without being incommoded by bad weather. The area in front is a smooth gravel plot, some feet below the level of the arcade, well supplied with garden chairs for the accommodation of the walkers.

"The building has three stories. The arcade is of the rusticated character. Above the arches, an elegant balustrade extends along the whole front and the ends of the fabric. Over the piers of the arcade arise fluted Doric pilasters, that support the architrave and cornice. The trygliphs of the former and the rich underpart of the latter have a beautiful appearance. The termination above the cornice is formed by another balustrade, that extends along the whole building. The front contains forty-two pilasters, and two tiers of windows above the arches, thirty-nine windows in each tier; to these add the lower windows, those in the ends, and in the back of the building, and there arises a total of three hundred and seventy-eight windows."— *Jewett's History of Buxton,* 1811.

The Square is connected with the Crescent by a colonnade; the colonnade extends along three sides of the Square; and the colonnade which skirts the internal area of the Crescent and the external area of the Square, forms a covered walk of a hundred and seventy-five yards in length.

The much extended ranges of baths are situated at the two extremities of the Crescent; that at the eastern end being devoted to the hot-baths, &c.; that at the western end to the natural baths, the wells for drinking the waters, &c.

Both these great ranges of buildings are covered, and their interiors lighted, by roofs of glass, arranged in the ridge and furrow form. Any undue degree of glare of light in the bath-apartments, where this might be objectionable, is controlled by internal blinds, which may be drawn at pleasure; and any additional ventilation, beyond that which is provided for in the ordinary way, is readily obtained by isolated and overlapping portions of the glass roofing, which are placed at intervals, in all needful positions, and which may be readily raised and lowered at pleasure. When the large amount of watery vapour necessarily discharged from the warm waters, as they are poured into the reservoirs, and thence into the baths, in such vast quantities, is considered, —and the considerable amount of heat which is given out from the waters, and the large quantity of nitrogen and carbonic acid gases constantly disengaged from them, are taken into account,—the great importance of a free ventilation at all times, and a command over the means of adding to its degree at pleasure, will be appreciated, and felt to be peculiarly needful. A dry air, of genial temperature, has been at length attained in the passages, dressing-rooms, and bath-rooms of both ranges of baths, by means of warming apparatus, of careful construction and efficient character. There is no particular in which the improvement in the new baths is more remarkable than in this. The bath apartments and passages had been thought to be unavoidably damp, owing to the constant current of the warm water through the baths. The complete manner in which this has been remedied, deserves to be pointedly mentioned.

The western or natural bath department, occupies a space of ground between the Crescent and the Hall, and has a comparatively small extent of frontage. This limited space is occupied by a handsome elevation of dressed stone, surmounted by a balustrade of enriched character, and presenting five compartments. Of these, the three in the centre are occupied by domed, semi-circular, recessed, and fluted spaces, of windowed size and shape; the base of every recessed space being formed of a vase, from the centre of which a jet of water may be made to play. *(See* the Illustration.) This architectural front has been adapted in its style to the Crescent with which it is connected, and to the uses of the building it appertains to; and it serves to illustrate very well, the suitableness of the stone of the adjoining gritstone formation for ornamental building,—its fine and beautiful grain, and the smooth surface and bold and sharp edges with which it may be finished and carved, either in relief or otherwise.

The elevation of the eastern, or hot-bath department, is not interfered with by any other building, and forms a decorated and substantial example of what must be called the Crystal-palace style of architecture,—a style which is one of the great creations of our times, and which is calculated to produce, directly or indirectly, a most important change in the character and details of modern architecture. *(See* the Illustration.) Presenting frontages of glass and iron, on the south and the east, which contain sixty enarchments; every enarched compartment having a breadth of four feet six inches; the building is nearly 30 yards in width, and more than 60 yards in depth. It is placed substantially on a base of wrought and smoothed stone.

Both these departments of baths are approached from the

colonnades of the Crescent and the Square by contiguous arcades; and there is a glass-roofed passage of communication from the Hall.

At the south-west corner of the Crescent, entered from the Crescent-colonnade, is the newly erected St. Anne's Well for the use of the drinkers of the water. This new well is on the site of the oldest St. Anne's Well that is on record, and close to the spot at which the spring emerges by which the well is supplied. The apartment containing this well is entered from the colonnade, without exposure to the weather. The well-room is lofty, and lighted from above; the well in the centre being surrounded by a ledge of marble, on which to place the glasses, — supported by a partition, from within which the water is dispensed to the drinkers.

On the north side of the entrance to the new St. Anne's Well, and close to it, is the entrance to the gentlemen's department of the natural baths; next to this is the entrance to the ladies' department of the natural baths; and next to this, and opening from the Crescent-colonnade in the same way, is the new well for the supply of the chalybeate water to the drinkers.

The size of the room containing the new chalybeate well, is twenty-two feet by sixteen feet, and lighted from above. The chalybeate water is poured from three orifices into an ornamental basin, in the centre of this apartment.

Every one of the baths in the natural-bath and the hot-bath departments, is separately supplied with the mineral waters, from closed reservoirs, in which the tepid waters are

collected, as they are poured out from the fissures in the limestone rock. The separate supply thus afforded to every one
of the baths, is so large, that the temperature of from 80°
to 82° is maintained, and that the gaseous and chemical
properties of the waters are preserved. In regard to the
natural-baths, there is not only this separate supply of the
mineral waters, but the waters are constantly running
into and out of them; the supply for every bath being
received directly from the reservoir which feeds it, and
carried away at once through the waste pipes into the river.
As in the instance of the other natural-baths, the baths
which are devoted to the use of the patients of the Buxton
Bath Charity, have likewise this independent, intact, and
abundant supply of the tepid waters.

The mineral waters are thus constantly pouring into and
out of every one of the natural baths, and in such quantity
that the temperature is always maintained at from 80°
to 82°; but the entire cleanliness of the several baths
is still more fully secured, by having them completely
emptied every night, and the sides and flooring duly scrubbed
and scoured with brooms and brushes.

The flow of the tepid waters is amply sufficient for every
purpose; and the amount of the waters which is discharged
altogether, is even considerably greater than the very large
quantity which is now made use of. Were the whole of
the flow of these mineral waters to be determinable,—
thus constantly discharged,—in unvarying quantity,—of the
unvarying temperature, at the moment of issue, of a

fraction of a degree more than 82° Fahrenheit,—and of un-changing chemical character,—the whole quantity poured forth would probably be found to be not less than 250 or 300 gallons per minute.

Dr. Short, writing in the year 1734, says, " all these four springs together," viz., those of the inner bath, the outer bath, St. Anne's Well, and and Bingham's Well, "throw forth in a year 97,681,860 gallons of water, besides the waste water that gets out of the bath, and the strong spring rising up in the middle of the bath level beyond St. Anne's Well, and the warm water which rises up in the hot and cold spring, and lastly the two small warm springs which rise up in the low ground, between the hot and cold spring, and the large spring in the sough, with several other oozings of warm water in sundry other places, the whole added together will be nearly double this computation." But even this computation, which Dr. Short states to have been the earliest which had ever been made, gives 185 wine-gallons, viz., 139 imperial gallons, per minute, as the flow of the four springs; which he was induced to estimate as only half that of the amount of tepid water actually dis-charged, if the whole had been collected, and none permitted to run to waste. The flow of Bingham's Well and of St. Anne's Well, according to Dr. Short's estimate, being deducted from the above, amounting to 26½ gallons per minute, would leave a flow of 112½ gallons per minute for the supply of the natural baths in the year 1734, i.e., forty-six years before the foundations of the Crescent were laid.

Fifty years after this estimate had been made by Dr. Short, an estimate of the flow supplying the natural baths, exclusive of that of the other wells and springs, was made by Dr. Pearson. This estimate was made in the year when the Crescent was completed, viz., 1784; and the flow is stated as having been "nearly 140 ale gallons per minute," which would be 116½ imperial gallons. These estimates confirm one another very remarkably, and justify our great confidence in the statements of these observers.

Much of this flow of the tepid waters supplying the natural baths, would seem to have been lost between the years 1784 and 1851; as, according to a report which was made to Mr. Smithers, by Mr. Eddy and Mr. Darlington, the engineers, in November, 1851, the amount of flow which supplied the natural baths at that time, was only 84¾ imperial gallons per minute. In the process of levelling and excavation for the formation of the new natural baths, a larger amount of flow has been regained than that which had thus gradually come to be wasted; and 129½ imperial gallons per minute of these tepid waters, are now poured forth for the supply of the natural baths exclusively, in addition to the flow by which the hot baths, and that by which the drinking wells are supplied. It will be observed, that this flow is greater to the extent of 17 gallons per minute, than the quantity of water supplying these baths in 1734; and greater by 13 gallons per minute, than the supply in 1784.

The depth of water in all the gentlemen's natural baths, is 4 feet 8 inches; and the depth of water in the ladies' natural baths, is 4 feet 2 inches. These baths are therefore used in the erect position, in order to admit of free exercise and movement during the period of immersion. This is essential in baths of water, at the natural temperature of the Buxton tepid springs: viz., 82°, Fahrenheit. Although the temperature of 82° constitutes a bath of tepid character, and may be said to be about 20° higher than the temperature of river water, in the summer season, in these high latitudes, it is nevertheless 16° below the temperature of the internal organs of the human body, and 13° to 14° below that of the surface of the body. A bath of 82° would therefore be unwisely made use of, in the recumbent position. The degree of muscular action which is involved in the maintenance of the body in the erect position, lessens the risk of chill attending or resulting from the use of the baths, even when the limbs are not kept in more or less active movement during the time of bathing. Crippled and paralysed conditions sometimes preclude any such movement of the limbs, or any very important amount of muscular exercise, from being had recourse to during the use of the bath. But, in most cases, active exercise is not thus precluded or interfered with, during immersion in the water; and the erect position in which the baths are used, leaves the trunk of the body and the limbs under full command, and renders every desired degree of exercise usually obtainable. The baths are of sufficient size, as well as sufficient depth, for this important purpose; and they are, moreover, surrounded with hand-rails, and supplied with swinging chains, in order that the bather may obtain any desired amount of exercise during the use of the bath.

It is by no means exclusively on account of the temperature of the water, that as much muscular exercise as is otherwise expedient and practicable, should be taken during the use of these mineral baths. The absorption of the water through the skin into the system, is indispensable to the effect of bathing in any mineral water. This absorption is secured and promoted by bodily exercise, and friction of the surface of the body, during the use of the bath. Very little absorption of the water is believed to take place through the skin, if the bather remains quiescent while immersed in the bath; and the greater the amount of friction of the skin, and the more active and general the degree of the muscular exertion which is made, the greater the amount of absorption under the same circumstances. This is applicable to baths of any temperature; but it is more particularly important in using baths of mineral waters; and more especially of those mineral waters, which may be chiefly dependent, for their medicinal action, upon the amount of the gaseous impregnation which they contain, —the absorption of such gas or gases being peculiarly under the influence of whatever promotes or otherwise the absorbent power of the skin.

All the baths are supplied with douches, or continuous jets of water, made to issue with a considerable amount of force, through nozzles of different sizes, and which may be directed against, and played upon, any part of the body, limbs, or joints, which may be more particularly affected. The douche is an exceedingly valuable remedy, in many chronic localised ailments. Sprains, and similar injuries of the textures near to the surface,—the seats of re-united fractures and reduced dislocations, which are often left for a long time after such injuries in a very imperfect and painful condition,—cases of spinal weakness, and localised chronic infirmities of rheumatic or gouty character,—and local forms of paralysis, sometimes traceable to exposure to cold and wet, sometimes to the effect of lead and other mineral poisons,—are found to derive much greater or more rapid effects from the use of the bath when combined with the douche, than when used without it. But there are many cases in which the bath cannot be justifiably made use of, to which the use of the douche is found to be applicable. When the circulation or respiration is much disturbed by some temporary derangement or permanent lesion, either in the heart, lungs, great vessels, or nervous centres, the propriety of immersion in any water, of any kind, or of any temperature, may be contra-indicated; and the use of the douche may often be tried with safety and good effect, for the relief of local ailments, or the lessening of general infirmities. It has to be stated, moreover, that so powerful are the effects of the mineral waters of Buxton, when used as baths, that it is found to be unwise to allow persons to use these baths every day. The use of the bath has to be interrupted, on the alternate, or on the third days, to diminish the risk of inducing undue stimulation in the first instance, or undue debility in the end. In many cases, the douche may be used without the bath, on these intervening days, which would otherwise be comparatively wasted. To meet such cases,—and they are necessarily very numerous, when both the classes of cases cited are reckoned,—douche-baths are provided, where any part of the body may be submitted to these water-jets without the use of the bath. It is impossible to attach too high a value to these douche-closets or douche-baths, in a medicinal point of view. The rapidity with which local lesions of chronic character are often relieved by means of the douches, the numbers of cases to which the douches alone are applicable, the use of the bath being contra-indicated, and the amount of time which may be saved by using the douches on the days when the baths are not used, justify the value thus assigned to these douche-baths.

The medicinal value of the douche is due to the greater degree of absorption of the mineral waters, through the skin of the parts submitted to its action; the effect of the impulse and percussion of the jet of water being tantamount to active friction with pressure. The readiness with which the degree of this friction may be controlled, by regulating the force of the jet and the time of the application; the much greater amount of this kind of friction that may usually be borne, without inconvenience at the time, or discomfort afterwards, than of rubbing with anything of harder character than the water itself; the amount of pressure with which the jet acts on the part submitted to it, answering the full purpose of most efficient shampooing; the perfect adaptation and equalisation of the pressure and friction over the whole surface douched, notwithstanding any curves or inequalities of the body or limbs, while the medicinal properties of the water are absorbed and brought to bear immediately upon the part or parts which may be more particularly affected,— are the evident reasons why the douches of the mineral waters prove to be of such great value in the treatment of many localised and disabling ailments. And it is not too much to say, that some of the most wonderful and gratifying instances of relief obtained from the use of the Buxton waters, have been referable to their use in the form of douches. A noble duke had his foot trodden upon by a horse. The foot was not apparently injured after the primary irritation occasioned by it had subsided. There was no perceptible swelling of the foot, nor thickening of the bones or ligaments of the arch of the foot, which had been injured. But there was much crippling, and some occasional pain. To walk was difficult; and to take an amount of walking exercise adequate to the wants and duties of life, was impossible. Months passed away; the most skilful surgical opinions and appliances were found to be useless. In three weeks, under the use of the baths and douches of the Buxton water, the patient was enabled to walk three miles continuously, without lameness at the time, or inconvenience afterwards. Such cases might be multiplied to any extent. This case is cited, because, from the high position of the sufferer, considerable attention was attracted to it at the time; and because it is one of many such cases,—of local injury without actual lesion, or irreparable damage, to the seat of such accident,—in which a cure by the use of the baths and douches of these mineral waters may be looked for with much confidence.

The gentlemen's department of the natural baths is entered by a corridor, which is sixty feet in length, and of an ample width and height; and which gives access to two large public baths, to the private baths, to the douche-baths, and shower-baths, of the water at the natural temperature.

The "Gentlemen's Public Bath, No. 1," or "Two-Shilling Bath," is contained in an apartment which is nearly fifty-one feet long, more than thirty-three feet wide, and upwards of twenty feet high, from the top of the water in the bath to the ceiling of the room. The bath itself is twenty-six feet in length, and eighteen feet in width. This large apartment contains suitable dressing closets, and all other desirable comforts and conveniences; and is lighted by means of a double tier of windows. This bath is on the site of the oldest of the baths; but the new bath is two feet longer, and two feet and a half wider, than the former one; the apartment is nearly double the height; it is well-lighted (instead of being somewhat dingily dark), and dry, and well warmed and ventilated (instead of being more or less close and damp at all times).

The "Gentlemen's Public Bath, No. 2," or "One-Shilling

Bath," likewise furnished with dressing closets and all other comforts and conveniences, is twenty-seven feet long, and fifteen feet wide. The apartment is not so lofty, nor so well lighted, as the No. 1 bath; but it is larger, and better lighted, than this bath used to be; and the area is well warmed and ventilated.

The gentlemen's private baths are eleven feet long, and five feet wide, with private dressing rooms, and every comfort and accommodation.

There are private douche-baths attached to the private baths, in which the douche may be used without immersion of the body; there being also douches in the immersion baths, in which both immersion and douching may be used. The douche-baths are likewise furnished with the shower-bath apparatus; so that the shower-bath may or may not be taken in connection with the douches, or the shower-bath be used without the douches, as may be desired.

The ladies' department of the natural baths is likewise entered by a separate corridor, sixty feet in length.

The Ladies' Public Bath is contained in an apartment which is thirty-nine feet long, and thirty-nine and a half feet wide. The bath itself is twenty-three feet long, and eighteen feet wide. There are dressing closets, and all desirable and comfortable appurtenances.

The Ladies' Private Baths are eleven feet long, and five feet wide, and supplied with douche apparatus. There are also private douche-rooms, where are douche-baths and shower-baths; either or both being used, as may be wished.

The Ladies' Private Baths are furnished with separate dressing rooms, and every accessory to comfort.

The baths of the waters at the natural temperature, provided for the use of the patients of the Buxton Bath Charity, are equal in every essential particular to those already mentioned.

The Men's Charity Bath is contained in an apartment which is twenty-six feet six inches long, and twenty feet wide; the bath itself being twenty feet long and fifteen feet wide. There are dressing boxes and every needful comfort, and a douche closet for the separate application of the douches without immersion.

The Women's Charity Bath apartment is thirty feet long and twenty feet wide; the bath being twenty feet long and fifteen feet wide. There are dressing boxes, douche-closet, &c.

Both these baths are lighted, warmed, ventilated, and supplied in every particular, as satisfactorily as the other baths.

It is mentioned by the late Dr. Joseph Denman, in a work entitled "Observations on the Buxton Water," published in 1801, in strong terms, as a great disadvantage to the usefulness of the Buxton waters, that no provision had been made for supplying baths of the mineral water at any higher degree of temperature than the natural heat. It was not until the year 1818, or seventeen years after the publication of this decided opinion in favour of warmer baths of the mineral waters, that this deficiency was in any degree supplied. But Dr. Denman could not have foreseen, nor could any adequate anticipation have been formed, as to the amount of benefit which would accrue from the use of artificially heated baths of the Buxton tepid waters, and the consequently greater and greater demand for these baths on the part of the public.

Much apprehension has been always entertained, lest the raising of the temperature of these waters, in ever so small a degree, might have the effect of impairing their medicinal qualities. Such an apprehension might seem to be the more justifiable, inasmuch as the opinion has come to be more and more generally held, that the medicinal effect. of the waters depends, to an important extent, upon the gases which they hold in solution, and which might be likely to be more and more driven off, as the temperature of the waters is more and more raised. It has to be remembered, however, that the whole of the waters poured forth from these springs, ·and supplied to the baths, have naturally the elevated temperature of 80° or upwards ; and that a very large proportion of the water in a bath is unmeddled with, until the moment of introducing the relatively small quantity of the same water heated, which is necessary to raise the water of the bath to such·higher temperature as may be required. Supposing the temperature of common spring or river water to be about 50°, a bath of 95° would require the addition of so much hot water as would elevate the temperature of the water 45° ; whereas, in the instance of the tepid waters of Buxton, the difference of temperature would be only 15°, and the addition

of one third only of the proportion of heated water to the
bath would be necessary. So small a proportion of heated
water has to be added to the natural water, to raise its temperature to that of any ordinary hot bath, that it has often been
impossible, when these baths have been in very great demand
from morning till night, to prepare a . bath in the hot-bath
department, at a lower temperature than 88°, or even sometimes than 90°; the heat of the marble sides and floorings
of the baths, and of the pipes conveying the hot water,
&c., being sufficient to raise by so much the temperature
of the natural water, without the addition of any heated
water at all. It may be justifiably advanced, that the
temperature of these mineral waters affords the greatest
facility for their use in the form of baths, at any required
degree of temperature, with the least possible risk
of impairing their effects. In a very large proportion of the
cases in which these baths are required, the natural
temperature is precisely that which would be desired. The
degree of heat is that, at which the slightest degree of shock
would be given on immersion, and a due amount of re-action
be rendered the most certain to follow the use of the bath,
—at which the good, without the evil effects of cold bathing,
would be experienced. At any higher temperature, the
regular use of the baths would be more likely to be
attended by debilitating effects. Whereas, as has been
stated, any such higher degree of heat for a bath may be
obtained most readily, by the addition of a very small proportion of heated water; and with so much less risk of

diminishing the amount of the medicinal properties of the waters. It may be advanced, that, if the Buxton waters had been of so much higher a degree of natural heat, that the water would have had to be lowered in its temperature, by the addition of cold or cooled water to it, in order to adapt its heat, for the purpose of bathing, to the requirements of a large number of invalids, more of the medicinal properties must have been diminished by such addition, or such exposure, than takes place under present circumstances; and if this water had been at such a natural degree of heat, that it could not have been used in any case without having been previously cooled by addition or exposure, the disadvantage and loss of properties must have been very important. But the proportion of invalids who use the natural baths, is greater than that of those who use the heated baths; and those who use the heated baths have, as nearly as may be, the full advantage of the medicinal properties of the waters, to the extent to which the water used in the baths is in its untouched and natural state; the bath being only affected in that proportion in which hot water is added, and in which the whole of the water in the bath may be supposed to be influenced by being mixed with so small a proportion of heated water, and by the temperature of all the water in the bath being raised any given number of degrees above the natural heat. The medicinal effect of the heated baths of the mineral water may have been hitherto further diminished, by the recumbent position in which they have had to be invariably made use of. No

amount of friction of the surface of the body by means of
brushes or hand-rubbing, has the same amount of influence
upon the absorbent powers of the skin, as active muscular
exercise of the trunk of the body and of the limbs ; involving
as this does, the friction of the skin with the water itself,
under the influence of the pressure exerted by the resistance of
the water to the muscular movements, and of the pressure due
to the weight of the superincumbent water in the bath itself.
Such a degree of exercise, as may be used in a bath which is
taken in the recumbent position, is necessarily much less in
degree than that which may be had in a deeper bath, taken in
the erect posture ; while the pressure of the water in the bath
on the surface of the body, is as much less as the bath is less
deep. So much greater an amount of effect as may be ascribed
to this cause, will be obtained from the large and deep baths
of the heated water, which are now constructed for the first
time.   The principal difference in effect between the heated
and the natural water, is however, in all probability, due to
the difference of temperature ; and this difference cannot be
so important as might have been supposed to be probable, for
the reason that so little of the water has to be made hot,
in order to raise the temperature of a bath of 82° to any
required degree of heat up to 95° ; beyond which temperature,
it is rarely found to be either needful or expedient in any
case, to raise the temperature of the water in these baths.
It follows, that the greater the extent to which the mineral
waters have to be heated, the greater the degree to which the
medicinal efficacy is diminished.   But when the above

statements are carefully considered, it must be admitted to
be wonderful that so small an addition of heated water to the
natural water as is required, should influence the medicinal
effects in any appreciable degree ; and the usual estimate,
that three baths of the water at the temperature of 95° are
only equal to two natural baths, is at all events as high an
estimate of the difference between the amount of relative
effect as is justifiable. And, accordingly, numbers of cases,
in which the use of these heated baths has to be trusted to
exclusively, the use of the colder natural bath being contra-
indicated by any individual circumstances of such cases, are
found to be relieved or cured, and to experience the well-
marked and specific effects of these mineral waters, as if the
natural baths had been made use of. Many periosteal,
neuralgic, spinal, paralytic, and atonic cases,—many cases of
rheumatism and gout, attended with much debility,—many
cases, in which acute or active morbid action has been recent,
or perhaps may have imperfectly subsided,—many cases, in
which disturbance or irritation of the heart's action, or of
the mucous or the fibrous tissues, or of any of the great
viscera, may render the shock of a bath of 82° inexpedient
or hazardous, and a less active agent than the unmodified
baths of the Buxton tepid waters to be preferable, whether
in the first instance, or throughout the course of the baths,
—such cases, and they are very numerous, find in these hot-
baths, adapted in temperature, &c., to the individual indications, the means of using these waters without risk, and with
every probability of benefit.

The hot-bath department, placed, as has been said, at the east end of the Crescent; occupying a frontage to the south of ninety feet, and to the east of 180 feet; is connected with the Crescent, the Square, the Hall, and the natural baths, by a colonnade; and is divided into two separate parts, one of which is devoted to ladies, and the other to gentlemen. None of the baths in this department have had to be placed beneath existing structures, as has had to be done in regard to two of the public baths in the other department; a sufficiently extensive and unoccupied space of ground has been covered throughout by a ridge-and-furrow roof of glass, and arranged internally in the best and most efficient manner.

The gentlemen's hot-bath department, to which a colonnade in the south front of the building, eighty feet in length, gives access, is entered by a corridor which is likewise eighty feet long. The several baths are entered from this corridor.

The Gentlemen's Public Hot Bath is contained in an apartment which is thirty-four feet long and twenty-six feet wide, and contains dressing-closets, the hot douche apparatus, and every desirable appurtenance. The bath itself, usually maintained at the temperature of 92°, is twenty-five and a half feet long and sixteen and a half feet wide, and is four feet eight inches deep.

The range of Private Baths is extensive and complete, with separate dressing-rooms, and with shower, vapour, and douche-rooms, and with every other appliance which

may conduce ·to comfort or advantage. These baths are
prepared of any heat that may be desired.

The private hot-baths are lined throughout with marble.
The other baths are floored with marble, the sides of the
baths being lined with the patent white porcelain-covered
bricks. The douches in the hot-baths, whether used in
connection with the baths, or separately in the adjoining
douche-rooms, are served at the required temperatures. The
private hot-baths are shallow, and used in the recumbent
position.

The ladies' hot-bath department corresponds exactly
with that appropriated to gentlemen. It is entered by its
separate corridor, eighty feet long, from the arcade on
the south front of the building; the corridor giving
access to a large warm public bath, twenty-five feet
long, and sixteen feet wide, and four feet two inches
deep, contained in a spacious apartment, with dressing-
closets, hot-douche apparatus, and all other comforts and
appliances; and the range of private hot-baths being likewise extensive and complete, with separate dressing-rooms,
—communicating with rooms for the use of the hot-
douche, hot-shower, or vapour-bath, without immersion in
the water.

This great building likewise contains the hot-baths for
the use of the patients of the Buxton Bath Charity. These
baths are approached by an entrance on the north of the
building. There are separate bath-rooms for men and for
women, each containing two baths, with dressing-closets,

I

douche-closet, &c., and entered from a comfortable waiting-room.

It had long been regretted that no other bath was attainable at Buxton than those of the mineral water, the use of which would be to many persons of no particular value, and which might prove to be unduly stimulating or otherwise injurious to such persons. People in health had long wished to have a bath at Buxton of ordinary cold water, in which they might bathe without doubt or apprehension of evil consequences from the medicinal character of the water. This wish has been attended to in the new ranges of baths. A large public cold plunging bath has been provided. This bath, supplied with water which has percolated the gritstone, is contained in an apartment which is forty-two feet long, and twenty-seven feet wide. The bath is lined with the white porcelain-covered bricks, and floored with marble. The bath is twenty-five feet six inches long, and fifteen feet six inches wide, and is provided with separate dressing-closets, and every comfortable appendage.

The account which has been given of the Baths of Buxton may be more clearly understood, and the character and amount of bathing accommodation be more completely appreciated, by reference to the subjoined ground-plans. The arrangement, the size, the number, and the convenient position of the baths will be thus understood, and the amount of provision which has been made for the wants of the public may be in some degree estimated. But the lightness, and elegance, and comfort of the interiors, and the happy adaptation of this new mode of architectural construction to these purposes, which have been obtained, must be seen and experienced to be adequately valued.

It will be seen at once, from an examination of the plans (*see* the two following pages), that the corridors of the ladies' and gentlemen's ranges of natural baths, are entered from the colonnades at the west end of the Crescent, as are likewise the apartments containing the St. Anne's Well and the Chalybeate Well; and that the corridors of the ladies' and gentlemen's ranges of hot baths, are entered from the colonnade at the east end of the Crescent. In order to realise their position and general effect, the natural-bath and hot-bath departments are to be considered to represent, respectively, west and east wings to the important structure with which they are connected, and with the architectural character of which they have been made to harmonise as far as possible. An attentive examination of the plans, in connection with a reference to the engravings at the beginning of the volume, will enable persons at a distance to understand the arrangements.

GROUND PLAN OF THE NATURAL BATHS.

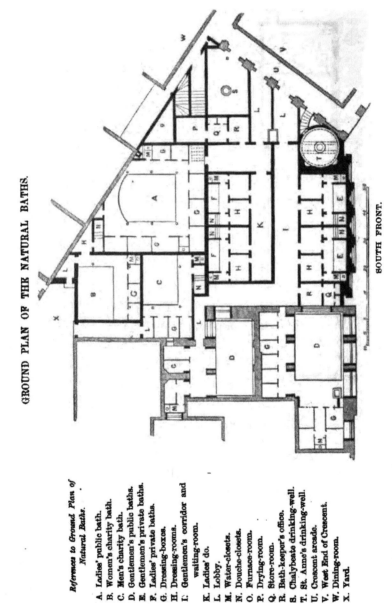

SOUTH FRONT.

References to Ground Plan of Natural Baths.

A. Ladies' public bath.
B. Women's charity bath.
C. Men's charity bath.
D. Gentlemen's public baths.
E. Gentlemen's private baths.
F. Ladies' private baths.
G. Dressing-boxes.
H. Dressing-rooms.
I. Gentlemen's corridor and
    waiting-room.
K. Ladies' do.
L. Lobby.
M. Water-closets.
N. Douche-closets.
O. Furnace-room.
P. Drying-room.
Q. Store-room.
R. Bath-keeper's office.
S. Chalybeate drinking-well.
T. St. Anne's drinking-well.
U. Crescent arcade.
V. West End of Crescent.
W. Dining-room.
X. Yard.

GROUND PLAN OF THE HOT BATHS, THE COLD SWIMMING BATH, &c.

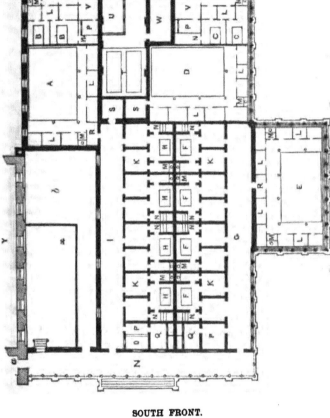

SOUTH FRONT.

References to Ground Plan of Hot Baths, &c.

A. Ladies' public bath.
B. Females' charity baths.
C. Males' charity baths.
D. Gentlemen's public bath.
E. Cold Swimming-bath.
F. Gentlemen's private baths.
G. Corridor and waiting-room.
H. Ladies' private baths.
I. Corridor and waiting-room.
K. Dressing-rooms.
L. Dressing-boxes.
M. Water-closets.
N. Douche-closets.
O. Ticket-office.
P. Bath-keeper's offices.
Q. Store-rooms.
R. Lobbies.
S. Drying-rooms.
T. Boiler-house.
U. Engine-room.
V. Waiting-rooms.
W. Coal-place.
X. Yard.
Y. East End of Crescent.
Z. Arcade.
a. Crescent arcade.
b. Billiard-room.

CHAPTER VI.

—•—

PRIMARY, SECONDARY, AND ALTERATIVE EFFECTS OF THE BUXTON TEPID WATERS. MORBID CONDITIONS FOR THE RELIEF OF WHICH THEY ARE USEFUL. CIRCUMSTANCES WHICH CONTRA-INDICATE THEIR USE. RULES FOR THE USE OF THE BATHS, AND FOR DRINKING THE TEPID WATERS.

THE effects of the baths of the Buxton tepid waters upon the human system, may be classed as primary, secondary, and alterative.

The primary effects include the shock immediately experienced on immersion in the water at the natural temperature, and the re-action which should immediately follow the shock; whether the degree of the re-action be no greater than is necessary to restore the general balance of the circulation, or the re-action continue during many hours afterwards; such an amount of re-action being included in the primary effects, although it may produce a very important degree of stimulation on the nervous and muscular systems. The secondary effects, which are seldom experienced until several baths have been taken,—or in any very important degree unless they have been taken during

several successive days, or unless the immersion has been
continued during an unusual length of time, or unless the
bath has been made use of by those suffering from special
morbid conditions,—do not show themselves until some
hours after the bath has been used; and are then characterised by excitement, at times amounting to a feverish state.
The alterative and ultimate effects of these baths, are not
usually produced until several baths have been taken; and
are then characterised by more or less of general depression
and languor, and their accompanying indications.

Under the head of the primary effects of these baths, is
the degree of the shock which attends their use at the natural
temperature; and this is believed to be so far peculiar, that
it is greater than would be occasioned by a bath of common water at this tepid degree of heat. This greater degree
of shock is probably due to the large proportion of gases
contained in these waters: a proportion which, as there are
$277\frac{1}{4}$ cubic inches of water in an imperial gallon, and $209\frac{1}{2}$
cubic inches of free carbonic acid and nitrogen gases in the
gallon of these waters, very nearly amounts to a charge of
the waters with their own bulk of gases. Usually, if the
bath has been used under proper circumstances, the shock
is of very short duration; and is followed, after a few
seconds, and during the immersion, by a vigorous re-action.
The degree of re-action is commonly greater than the
degree of the shock which had preceded it, and much
greater than would attend the use of a bath of ordinary
water at the same temperature. The re-action, involving a

general glow of heat over the whole body and limbs, a somewhat accelerated circulation, and a remarkable buoyancy of the feelings, usually continues during several hours after the bath, and is attended by an increase of appetite, and a marked degree of excitement of the spirits. The special effect of the baths of these mineral waters, in regard to the primary shock, and the re-action which immediately follows it, is however chiefly one of degree; the same effect, but less in its amount, being commonly experienced after bathing in common water. The warm baths of the Buxton mineral waters produce necessarily less shock, in proportion as the water of the bath is more nearly of the same temperature as that of the human body; and the degree of the re-action is usually less, other things being equal, than that which follows the use of the natural baths; but it is usually greater, to a very marked extent, than commonly follows the use of a warm bath of ordinary water.

The secondary effects of the Buxton baths are of a still more peculiar character; differing much more from that which follows the use of a bath of ordinary water, at any temperature. Seldom manifested until from eight to sixteen or eighteen hours after the use of the bath; and seldom, unless under specially excitable and morbid conditions, until after several baths have been made use of; the degree of the secondary effects is added to, by the longer time that the bather has remained in the bath, by the more frequent repetition of the bath, and by the morbid or the constitutional susceptibility of the system to become unduly stimulated. These secondary effects include the increase of gouty and rheumatic pains, so commonly experienced at the commencement of a course of these baths, in such conditions of system; and likewise the thirst, restlessness, loss of sleep, feverish symptoms, and less active state of the excreting organs, during the course, and especially during the earlier part of the course, of the baths; for, as the course of baths is continued longer and longer, the secondary effects gradually subside, and are succeeded after a longer or shorter interval, which varies much in different cases, by the alterative effects; the feverish condition sometimes subsiding altogether; but in most cases, in the first instance, alternating with the indications of the alterative action.— alternating with the depression, languor, and eventual debility, which mark the full alterative action of these mineral waters.

The alterative action of the baths is then essentially characterised by symptoms of debility; and such symptoms —apart, of course, from the relief of pain, the partially or entirely restored use of crippled limbs, or the generally improved state of the several organic functions—constitute the great and conclusive proof of the full medicinal action of the baths of the Buxton tepid waters. And it may be strongly said, that the full effects of the baths can seldom be considered to have been obtained, until some indications of diminished general power have been shown. It is often of some importance, in cases where much morbid condition has to be removed, to determine how far the alterative action of the baths should be continued; and in most such cases, the course has to be interrupted and resumed many times. Sometimes two courses, and more rarely three courses of the baths, may be advisably used in the course of the same year; interrupted by a sufficient interval of time, to enable the powers and general balance of the system to be so far restored, as to justify the resumed use of so powerful an alterative as these waters are. It is often a great mistake to infer, that the full benefit derivable from their use is certain to be obtained from one, or two, or even more courses of the baths. I have seen—in the course of the eighteen years, during which the effects of these waters have been under my observation—very many cases in which a large and even unlooked for degree of benefit has been obtained; but in which, after a persevering use of the baths through two or three courses, they have been abandoned; because they had not proved to be completely curative, although they had been so largely remedial. I have seen the bed-ridden sufferers from gout and from rheumatism, enabled to walk about again, with the aid of crutch, or stick, or both; and enabled, contrary to any reasonable expectation that could have been entertained, to resume the horse-exercise, which had been impracticable for months or years; and yet the hopeful continued use of these waters has been abandoned, because a more rapid and entire relief has not been afforded. It cannot be too early, or too anxiously, impressed on the minds of such sufferers from the most severe forms of gout or of rheumatism, that a strong constitutional bias, the gradual result of hereditary predisposition, or of the misdoings of years, and unchecked morbid action, may not be entirely curable, even by the use of such a great agent as these waters; or if curable, may only prove to be so after a long, persevering, and patient use; and that a certain amount of relief, the degree of which may be even beyond all reasonable hope, ought to afford an amply sufficient motive for their continued use.

The diseases for the relief of which the Buxton baths are found to be the most eminently useful, are, rheumatism, gout, neuralgia, and certain forms of spinal, uterine, and dyspeptic affections. Many of the disordered conditions which are incidental to old age—much of the deranged health incidental to middle age in females—much of the uterine irregularity and disturbed condition, incidental to females at various periods of life—much of the nervous weakness, that is indicated by tic-douloureux in its various forms, sciatica, &c.—much of the functional derangement of the kidneys, which is consequent upon exposure, intemperance, or advanced life—much of the disordered and painful conditions of the bladder, &c., dependent on old age, gout, &c.,—much of the local loss of nervous, and thence of muscular power, dependent upon the poisonous effects of lead, mercury, &c.—are usually remediable, and in an important degree, by the use of these mineral baths. The painful or crippling consequences, which often follow such injuries as fractures, dislocations, sprains, bruises of tendons and ligaments, and the like, are commonly influenced and

relieved by the use of these baths, in the most satisfactory
degree.

The presence of acute inflammation, and the existence of
organic disease in any of the great organs essential to life,
usually contra-indicate altogether the use of these baths.
The intimate connection between paralysis and disease of
the brain, or spinal chord, or their immediate envelopements,
—the equally intimate connection between rheumatism and
affections of the heart,—the frequent occurrence of pains of
rheumatic character, in connection with the general derangement of health, consequent upon affections of the liver or
the kidneys,—the connection between gout and visceral
congestion, and all the important bearings and consequences
of such congestion,—the degree of liability there is, in acute
or subacute gouty and rheumatic states, to metastasis,—are,
severally, cogent reasons for the exercise of care and judgment in the use of the baths.   It cannot be said too
strongly, that no invalid should leave his home in order to
make use of these baths, without the express advice and
sanction of his usual medical attendant; and medical men
cannot be made too fully cognizant of the stimulating and
alterative character of these mineral waters.   And, moreover, as no medical man who has not personally been
concerned in the use and effects of the waters, can know so
much about them as those who have their effects under
their continual observation, it may be said, with equal truth
and emphasis, that no person ought to use these waters
without the sanction and direction of a medical man resident

in Buxton. It is my duty to state this in so many words, and to urge it upon public attention; and the seniority of my position enables me to do this with a less chance of misconstruction, and justifies me in doing so.

The warmer baths of the Buxton waters are weaker, less stimulating, and less alterative in their effect, in the same proportion as they are raised in their temperature above that of 82°; and inasmuch as, when not heated beyond the temperature of 95°—that is, when no more of the same water heated is added to a much larger bulk of the water at the natural temperature, than is required to raise the whole water in a bath to 95°, or less—not only a sufficiently large, but a definable proportion of the medicinal effects, is retained; inasmuch as the shock, with its risks, in the cases of feeble and excitable invalids, is thus got rid of; inasmuch as the primary stimulation, and the secondary febrile state, and the ultimate alterative effects, are thus modified; the use of these warmer baths frequently serves as a valuable introduction to the after use of the natural baths, and as a very useful substitute for them in cases where debility, or excitability, would render the use of the natural baths unwise or unsafe.

Neither the natural baths nor the warmer baths ought to be used every day. When so used, the alterative effects are very liable to be manifested suddenly, and in excess. Strong men who have ventured to bathe in these waters every day, have often become suddenly and very unnecessarily debilitated; and in the case of rheumatism and other ailments having been the occasion for using these baths, the

disadvantage has arisen that the bathing has had to be discontinued for a time, and sometimes for months; the full,
and otherwise realisable degree of relief, not having been
obtained. The impatience to secure the benefit from the
use of the baths in as short a time as possible, and the
anxiety to return to their homes and occupations, lead many
to make such excessive use of them, and supply ample
and conclusive experience as to their powerful character.
Generally the baths are to be used every other day; or
on two successive days, their use being omitted on the
third day.

The time of the day for bathing is a question of much
importance. The baths have usually most effect when
used before breakfast, and are commonly the best borne
about three hours after breakfast. The system is found
to be more susceptible to the action of medicinal agents,
before breakfast; probably because the nervous and vascular
powers are more vigorous at that time, and the tissues are
then in a more absorbent condition. It is well known that
stimulants produce greater effect when taken before breakfast, and tonics are seldom wisely directed to be taken
at that time; and, on the same principle, the baths of
these mineral and stimulating waters are not by any means
always wisely ordered to be used before breakfast; and,
on the contrary, many invalids are unable to bathe with
comfort or even safety before breakfast, who do so without
discomfort between breakfast and dinner. The time for
using the baths after breakfast, depends much upon the kind

and amount of the breakfast, and the rapidity of digestion.
The larger and heavier the meal, and the more slow the
digestion, the longer should the bath be deferred. And, on
the other hand, in the cases of much enfeebled persons,
and of those who from habit, or want of appetite, or
general debility, eat but sparingly at breakfast, and of the
most easily assimilated kinds of food, the use of the baths
may not be wisely deferred beyond the end of the second
hour after breakfast, as the complete digestion of the food
in the stomach is apt to be followed by an unmistakeable
degree of languor, and the bath is not so well borne under
such circumstances. It is generally advisable to use the
baths between breakfast and dinner, at the commencement of the course; and in some cases to use them before
breakfast afterwards, when the effects shall have been
proved to be moderate, and the extra effect obtained by
bathing before breakfast may even be thought to be
desirable. Usually, more baths have to be taken during a
course, if the baths are used after breakfast; and a longer
period of immersion should be ordered if the baths are
used after breakfast, than if used before breakfast. The
more feeble the person, the more excitable the individual
constitution, and the more febrile or inflammatory the
nature of the ailment, the less desirable it is to bathe
before breakfast; and the less susceptible the individual
system, the stronger the system, and the more obstinate
and unimpressible the nature of the ailment, the more
desirable it is that the baths should be used before

breakfast. The time during which persons should remain
in the baths varies very much; but should seldom if ever
equal the long periods which are found to be expedient, in
using many of the continental mineral baths.   From one
minute, or less, to ten minutes, and very rarely to fifteen or
twenty minutes, is usually found to be a sufficient time for
the immersions in the natural baths; and from three minutes
to twenty minutes, and very rarely longer than the latter
time, in the warmer baths.   Generally, the time of immersion should be longer in the warmer baths, than in the
natural baths; and longer, the higher the temperature
of the warmer baths may be.   Usually, if the temperature
of the warmer baths is not gradually reduced, in order to
bring the case more and more within the influence of the
medicinal effects, or in order to prepare the patient to com-
mence the use of the natural baths, the time for remaining in
the baths should be increased.   It is seldom wise to begin
the course of either the warmer or the natural baths, by
remaining in the water for the same time which may
properly be allowed after several baths have been taken, and
the system has become somewhat accustomed to the stimulating effects of the waters.

   When the warmer baths are not indicated,—either as
preparatory to the use of the natural baths, or as being
exclusively adapted to the individual case,—and the use of the
natural bath might be justifiably begun at once, the use of
the warmer baths may be deprecated as involving exactly as
much loss of time, as these baths are less powerful in their

effects than the natural baths, besides being in some instances less suited, or even occasionally altogether unsuitable. This often applies to dyspeptic conditions,—to cases of gout, rheumatism, &c., attended by cutaneous irritation, with or without the scorbutic character,—to relaxed states of the female constitution, without visceral congestion or obstruction,—and to such cases of spinal derangement as are not marked by irritation, but by general or local relaxation, and consequent diminution of power and defective function. There are many cases of hysteria, chorea, facial tic, &c., in which the warmer baths are either used without advantage, or even disadvantageously ; and in which the use of the natural baths is attended with the best effects.

But there is a much larger number of morbid conditions, in which the use of the natural baths is unsafe or unjustifiable, and in which the use of the warmer baths is beneficial. Such are cases of gout, rheumatism, or neuralgia, accompanied by marked irritability of the general system, or by an evident tendency to assume an inflammatory character. Such are cases in which there is irritation or disturbance of the heart's action ; cases in which there is a congestive or irritative condition of the mucous membranes ; cases of hepatic congestion, irritation, sluggishness, or disturbance of function ; and, generally, cases of congestion or irritation of any of the great internal organs ; such congestion or irritation not being sufficient in degree, or in the importance of its bearings or connections, to justify the withholding the baths when otherwise strongly indicated for the relief of the special conditions in which these baths are useful.

The baths should not be made use of after dinner, and not later than three or four hours after breakfast, if the system is very excitable, or there is any reason to infer a probability that the baths may disagree. The most convenient times for using the baths, when not thus contra-indicated at those times, are before breakfast, and an hour before dinner. It is necessary, in almost all cases, to return to the lodging as soon as possible after leaving the bath, and to remain within doors during two or more hours afterwards, according to the season and the weather. In dry and warm weather, two hours will generally be sufficient for this purpose; but in cold and damp weather, it may be advisable to remain in the house three or four hours, or even during the whole remainder of the day. Remaining thus quiet during so long a time after using the baths, not only diminishes the risk of taking cold after them—for the degree of excitement produced by the baths renders this unlikely, and of rare occurrence—but it lessens the chance of undue excitement of the system after the bath, and is therefore to be generally and strongly advised. The tendency to go to sleep after using the bath, which is often great, should be resisted in all cases. Sleeping, until some hours after using the baths, almost always deranges, and excites, and adds to the risk of congestion, and should be watchfully avoided.

Drinking these waters produces much the same effect on

the system, as is produced by the baths; they are, however, more immediately stimulating, and less eventually alterative in their effects, than the baths.   They act occasionally as an aperient; but this is uncertain and unsatisfactory, and rather indicates irritation than relief, suggesting their discontinuance until some corrective medicine has been made use of, such as may relieve the congestive or morbid condition of the abdominal organs and secretions, upon which this effect from the internal use of the waters almost always depends.   They ought to act upon the kidneys; and their value as a diuretic, and corrective of some morbid conditions of the urine, is very great.   They ought not to occasion headache, or thirst, or loss of appetite, or broken or disturbed sleep, or feverishness; but they ought, on the contrary, to promote appetite and digestion.   Their use is often of great value in such cases as the baths would be prescribed for, but from circumstances contra-indicating their use.   The internal use of the waters is eminently auxiliary to the effect of the baths, and should always be used in connection with the use of the baths, when not contra-indicated.   The degree of excitement immediately occasioned by them, however, makes it needful to forbid their use in many cases, in which the baths may be used without any disadvantage.   This remark applies to a large number of cases of gout.   The waters should not be drunk either immediately before, or immediately after the use of the bath.   It is desirable that the possibility of the twofold excitement, caused by using the waters in both ways, without some interval of time,

should be avoided. If the bath is not used before breakfast, the first glass of the waters, and in some cases
two glasses of the waters, may be taken before breakfast; and the second or third glass may be taken, when not
interfering with the time of the bath, three hours after
breakfast. Or the second, or third, or fourth glass of the
waters, may be taken two or three hours after the bath, or
an hour before dinner, or three hours after an early dinner.
It is seldom necessary to take more than two half-pints of
the waters every day. There are three differently-sized
glasses at the St. Anne's well; the one containing a quarter
of a pint, the second one-third of a pint, and the largest
half a pint. It is generally advisable that the smallest glass
should be used at the beginning of the course; increasing
the dose again and again, at the interval of a day or two, if
no contra-indication should occur. The waters are so fully
charged with gas, and until the system has become accustomed to its use, the gas is so apt to occasion some degree
of giddiness or even headache, that it is prudent at first to
drink the waters by sips, and even to hold the glass in the
hand a few seconds before beginning to drink it; but this
seldom applies to more than the few first times of drinking
the waters; and afterwards it is desirable, in order to secure
the whole of the gas, or as much of it as may be, to drink
the waters as rapidly and immediately as possible, after
receiving it from the attendant at the well. It is desirable,
if possible, to walk after drinking the water, in order to
expedite its absorption from the stomach. The amount of

walking found to be desirable on this account, varies from ten minutes to an hour, or more. The expediency of remaining within doors, during one or more hours after using the baths, and of walking for some little time immediately after drinking the waters, is an additional reason why the use of the baths and the drinking of the waters should, if possible, be at different times of the day. The waters should seldom be drunk later than four or five o'clock in the afternoon. When more than two glasses of the waters during the day are indicated, the additional quantities may be taken from half an hour to an hour or more after the previous glass, according to the readiness with which the waters are found to be absorbed from the stomach. The internal use of these waters is often extremely useful, in cases of bronchial irritation and relaxation, urinary affections, and irritation of the bladder, in which the use of the baths may or may not be advisable.

CHAPTER VII.

— ◆ —

ANALYSIS, CHARACTER, AND USES OF THE CHALYBEATE WATER.

BETWEEN the limestone and gritstone formations at Buxton, there is a narrow bed of shale, containing a considerable proportion of iron; and from this spring arises a very useful chalybeate. This water has been long and extensively used and valued. It is an excellent chalybeate; fortunately, as nearly as may be, free from alum. It is therefore as little astringent in its effects as possible, and acts as a mild and efficacious tonic, producing the usual effects of iron upon the constitution, and in the perfectly satisfactory way that only chalybeate waters attain—being more certain in the effect, more secure of absorption, and less apt to heat the system, or engorge the membranes or viscera, than any artificial way of exhibiting iron medicinally. This water has a distinct chalybeate taste, is colourless and inodorous, and is of the ordinary temperature of the atmosphere. It was analysed by Dr. Lyon Playfair, in the year 1852, with the following results. The imperial gallon was found to contain—

	Grains.
Proto-carbonate of Iron	1·044
Silica	1·160
Alumina	trace
Sulphate of Lime	2·483
Sulphate of Magnesia	0·431
Carbonate of Magnesia	0·303
Sulphate of Potash	0·147
Chloride of Sodium	1·054
Chloride of Potassium	0·460
	7·082

The value of such an excellent and simple chalybeate water is necessarily very great, in a place which is resorted to by such large numbers of invalids, for the use of the baths of the tepid waters, or for the benefit of the great and useful change of air, which the mountain elevation, and the limestone and gritstone soils of the Buxton district, afford to important classes of invalided conditions. There are great numbers of invalids who use the Buxton baths advantageously, in whose cases the internal use of the tepid waters is not advantageous, and in whom the internal use of the chalybeate water proves to be eminently serviceable. There are many cases in which the use of the tepid waters is not indicated, either as baths or internally, that take the chalybeate water with advantage; and there are some cases, of not unfrequent occurrence, in which the chalybeate and tepid waters are mixed together, and so taken with much benefit. It is curious, that when so mixed, they are often found to have a laxative effect; whereas the tendency of the chalybeate water alone, and in some degree the tendency of the tepid water, is to produce a constipated rather than a laxative effect.

The chalybeate water should not be taken before breakfast. When so taken, it is liable to induce headache, and feverishness, and gastric disturbance, in cases where it suits well when taken at other times. The best times for drinking the chalybeate water, are between the breakfast and the dinner, when two or more glasses, of larger or smaller size, may be taken at intervals of one or two hours. The quantity taken at a time may be from a quarter of a pint to half a pint. It should not occasion headache or feverishness ; it should promote appetite and digestion, and not interfere disadvantageously with either. If it should occasion thirst, or uneasiness, or sense of distension of stomach, it should probably be discontinued. It is advisable, if possible, to walk after drinking the water.

The chalybeate water is of much value as a collyrium, in many cases of weakness and chronic irritation of the eyes. It is of more use as an eye-water, than might have been expected from a reference to its composition. It should be applied by means of the usual eye-glass adapted to this purpose ; and it should be used freely.

The chalybeate water is extremely valuable as an application, auxiliary to the baths and douches of the tepid waters, for the relief of indolent swellings of the joints, &c., so often left after the subsidence of rheumatic conditions. When poured over the joints affected, in directed and regulated quantities, once, or twice, or three times a day, it has appeared to stimulate the absorbents to a much greater degree, than has been found to attend the similar application of common water at the same temperature. This success has led to its use by means of sponging and friction, in similar cases, and with good effect; and this, to the use of the affusion of this water down the spine, and to sponging and friction of the back with it, in some forms of spinal weakness and irritation, with good effect. In some paralytic conditions, likewise, whether dependent on cerebral or spinal affection, the use of the affusion of this water, or of sponging and friction with it, over the head, or down the back, according to the nature of the case, has often been attended with good effect.

CHAPTER VIII.

THE SUPPLY OF GRITSTONE WATER FOR DOMESTIC AND ORDINARY PURPOSES.

THERE are not many questions which are of greater importance in regard to the health of people in different localities, or which have met with less full and practical consideration, than the extent and the character of the water supplied for domestic and ordinary purposes. Use has been made of the advantageous position of Buxton, in regard to this important particular. There is an abundant supply of remarkably pure water. The water of calcareous districts generally is more or less objectionable for domestic uses, inasmuch as, according to its degree of hardness, the solvent power of the water is lessened; and all cleansing and washing purposes so far interfered with, and the infusion of tea, the boiling of vegetables, &c., so much less readily completed. And, moreover, even persons who reside habitually in calcareous districts, are occasionally found to suffer from the calcareous matters contained in the water, and more especially from their astringent effect; and strangers, not accustomed to the use of such waters, suffer much more frequently, and to a greater degree. It is one of the valu-- able circumstances referable to the gritstone formation which adjoins Buxton, that the water-supply for domestic purposes is derived from it. The water is brought from two sources : the one, of higher level, locally known as the cold springs, arising near to Comb's Moss, on the right of the Manchester road, about a mile from Buxton; the other, from a stream in the lower grounds, nearly a mile to the left of this place. The waters from both these places were examined by Dr. Lyon Playfair, with the following results :

"*Buxton*, 30*th August*, 1852.

" DEAR DR. ROBERTSON,

" I have examined the water with which Buxton is now supplied (for domestic and ordinary purposes), and find it to be pure and soft, such as is indeed to be expected from a water flowing from the millstone grit. Its hardness is of two degrees ; that is to say, it is of the same hardness as would be given to one gallon of distilled water, by dissolving in it two grains of chalk, (carbonate of lime).

"The water from the brook, which is intended to be used for the further supply of Buxton, is 4·35° (4½ degrees) of hardness. This is also a soft water, though twice as hard as the previous sample.

"Perhaps you may judge better of the relative qualities of these waters, by contrasting them with the water of the river Thames, which is about 13 degrees of hardness.

" I am, very sincerely yours,

" LYON PLAYFAIR."

A curious circumstance connected with the use of this water by strangers is, that it is occasionally found to be too pure. The taste of the water is considered to be vapid by

x 2

some, who have been accustomed to use waters impregnated strongly with earthy or saline matters; and the change to the use of this water, from that of less pure water, has been known to produce some degree of discomfort in very delicate and highly sensitive systems.   In such cases, the water of the cold calcareous springs has been desirably substituted for the gritstone water; and Buxton is variously and sufficiently supplied with water of this kind.   The brook which runs across the fields from Wye-head, is of this kind of water.

CHAPTER IX.

—◆—

HISTORY, PROGRESS, POSITION, AND USEFULNESS OF THE BUXTON BATH CHARITY.

It is a peculiarly grateful task, in an account of Buxton and its mineral waters, to set forth the doings and claims of the Buxton Bath Charity. This charitable institution has gradually arisen, from apparently small beginnings, to its present position of extensive usefulness. It is recorded by Dr. Jones, in his work so often cited, that Buxton was much resorted to by poor afflicted persons, in the middle of the sixteenth century; and indeed the petition which has been previously mentioned, from the inhabitants of the adjoining village of Fairfield, which was addressed to Queen Elizabeth, in the year 1595, for a grant from the Royal bounty for the maintenance of a chaplain, on the ground that they were too much impoverished to do this for themselves, by reason of the urgent and continual claims for relief on the part of the many poor sufferers, resorting to the Buxton baths, confirms the opinion, that the numbers of the poor persons who then made use of the waters, must have been very considerable. It may be fairly gathered from the various ancient writers on these waters, that these poor persons had been so far aided, from time immemorial, as to have had the use of the baths allowed to them gratuitously. Dr. Jones, indeed, speaks of " the treasury of the bath " as being partly devoted " to the use of the poor that only for help do come hither;" but whether this is said in furtherance of a fund that had been already in existence, or only by way of suggestion for the formation of such a fund, does not clearly appear. Dr. Jones adds, however, an appeal that deserves to be quoted in all works that may be written concerning the waters of Buxton:—" If any think this magisterial imposing on people's pockets, let them consider their abilities, and the sick poor's necessities, and think whether they do not in idle pastimes throw away in vain twice as much yearly; it may entail the blessing of them that are ready to perish upon you, and will afford a pleasant after reflection. God has given you physic for nothing; let the poor and afflicted (it may be members of Christ) have a little of your money; it may be the better for your own health. Heaven might have put them in your room, and you in theirs; then a supply would have been acceptable to you."

It would seem, then, that the invalided poor, resorting to Buxton for the use of its baths and waters, had thus early met with aid as to their gratuitous use, and also with a kindly solicitude and attention from medical men in attendance at the baths; for it would appear to have been the custom, in earlier times, as it is said to be even now in some of the continental watering places, that the medical attendants

were wont to afford their instructions in the bath-apartments,
and personally superintend the fulfilment of their directions.
So much is left in obscurity, as to periods of time less
remote from our own than that now referred to, that it is
little wonder the early history of Buxton should have left no
more traces than these of the origin of the Buxton Bath
Charity; and it cannot but be regretted, that the ancient
records, which appear to have been kept at these baths, by
the medical attendant when he was present, and by the bath-
keeper in the absence of the physician, of the "name, place
of abode, coming thither and departure, condition, calling,
disease, and symptoms, and the benefit received," should have
been destroyed. A more curious and interesting record than
such an ancient register would now be considered to be,
cannot well be imagined. It is not, however, known, at
what distant period of time, baths were first provided at
Buxton for the gratuitous use of the poor; nor is it even
known, when baths were first provided for the separate and
exclusive use of the poor bathers.

I am in possession of a printed document, which bears the
date of 1785, in which it is stated that a pecuniary fund for
the assistance of poor bathers had originated in the year
1779; and this paper is evidently a copy of an annual report
of this charity, which has probably been issued regularly
from that time to the present: although I have not been
able to meet with any copies of these reports between the
years 1785 and 1818. It is not known when "the poor's
treasury," if such really had existed, ceased to be supplied

with funds; but enough has been said to show, that the
Buxton Bath Charity may claim to be one of the most
ancient charitable institutions in the kingdom. The fund,
which has, at all events, since the year 1779, been regularly
provided by charitable persons, for the assistance of poor
bathers, must have been at first of a very moderate amount,
as the number who were to receive pecuniary relief from it was
limited to "sixteen objects at one time," and it was only given
during the six summer months. Since the year 1820, the
annual reports of the institution show that 33,709 patients
have been admitted to the use of the baths connected with
it; and that of this number, 27,008 were dismissed as having
been either cured or much relieved, only 6,701 having had
to be sent away as being only somewhat relieved, or no better,
from the use of the waters. Extending over so many years
as from 1820 to 1853, and embracing such large numbers
of cases, these records are ample enough to satisfy the most
sceptical inquirer. And as to the nature of the ailments, in
regard to which so great a degree of success has been
obtained by the skilled use of the Buxton waters, it should
be stated that a very large proportion of the cases have been
those of rheumatism,—and of rheumatism, for the relief of
which hospital and dispensary appliances, and the efforts of
the private practitioner, have been tried in vain; and,
moreover, that the results have been obtained, for the most
part, by the use of the Buxton baths and waters for average
periods of only from three to four weeks. The chronic and
obstinate character of the generality of these cases, is

sufficiently explained by the difficulty with which poor
people are enabled to leave their homes; and by the fact,
that, notwithstanding the pecuniary assistance afforded by
the institution, and the gratuitous provision of medical
advice, medicines, and baths, few poor persons are enabled
to defray the cost of a journey to Buxton and back, and
remain there during the three or four weeks necessary to
complete a course of the baths, at a less outlay from their
own resources, than about two pounds sterling. This circumstance, moreover, causes attempts to impose upon public
liberality on the part of healthy persons, or of persons
suffering from less severe or less obstinate forms of ailment,
to be too unprofitable, to be of frequent occurrence.
That many much afflicted poor persons should, notwithstanding so large an expenditure, from scanty funds, contrive
to visit Buxton every year, or almost every year, as the only
means which they find to be effectual, in warding off such a
degree of crippling of the limbs, as would prevent them from
pursuing the occupation by which they earn their livelihood,
may be regarded as supplying strong evidence of the
medicinal value of the baths and waters.

The following is an abstract of the Report of the Buxton Bath Charity, for the year ending September 5, 1853.

BUXTON BATH CHARITY,

Instituted for the relief of poor persons, from all parts, suffering from Rheumatism, Gout, Sciatica, and Neuralgia; Pains, Weakness, or Contractions of Joints or Limbs, arising from these Diseases, or from Sprains, Dislocations, Fractures, or other local injuries; Chronic forms of Paralysis, Dropped Hands, and other Poisonous effects of Lead, Mercury, or other Minerals; Spinal Affections, Dyspeptic Complaints, Uterine Obstructions, &c.

SUPPORTED BY VOLUNTARY CONTRIBUTIONS.

1334 patients have been admitted to the baths, during the year ending on the 5th of September, 1853 ; of whom were—

Cured, or much relieved 710
Relieved 519
No better 41
Remaining on the Books 64
———
1334

In addition to the new charity baths supplied with the mineral waters at their natural temperature, which were erected last year, the trustees have now the satisfaction to report the completion of new hot baths for the use of this charity, adjoining the new public hot baths recently erected for the Duke of Devonshire; and they cannot omit again to acknowledge his Grace's great liberality, in aiding to provide, in so efficient a manner, for the relief of the poor resorting to the baths at Buxton. The medical trustees hope, by means of these baths, to extend the benefits of the institution, in a considerable degree. Hot baths, in connection with this charity, are now available for a much larger number of patients than heretofore; and a greater range of disordered conditions is brought within the advantageous use of the mineral waters.

RULES.

1. A Donation of £10 constitutes a Life Subscriber, with power to recommend a patient to the full benefit of the charity, as per Rule 2.
2. A Subscriber of One Guinea may recommend a poor patient, who will receive medical advice, medicines, the use of the baths, and a gratuity of five shillings weekly, for the period of three weeks.
3. A Subscriber of Half-a-Guinea may recommend a poor patient, who will receive medical advice, medicines, the use of the baths, and half-a-crown weekly, for the period of three weeks.
4. A Subscriber of Half-a-Crown may recommend a poor patient, who will receive medical advice, medicines, and the use of the baths, without pecuniary assistance.
5. Any person in need of the use of the charity baths is admitted thereto, on producing a certificate from the clergyman of his or her parish, or from his or her medical attendant, as to inability to pay for the same.

6. No person is allowed to use the charity baths without the written sanction of one of the medical trustees of the Institution.

7. In consequence of several inappropriate cases having been sent, there must be annexed to every recommendation a medical certificate of the nature of the complaint, and of the patient's fitness for the use of the baths, without which certificate no patient can in future be admitted.

All applications in reference to the admission of patients, are to be addressed to the Secretary, Mr. JAMES WARDLEY, Buxton, Derbyshire.

The fifth of these Rules is so liberal in its character, that great care should be used in ascertaining, as far as may be, the poverty of the persons recommended. The neglect of the seventh of these Rules so frequently occasions disappointment, and a loss of the time, trouble, and cost of the journey to Buxton, that the Rule cannot be too earnestly pointed out to the attention of medical men, and of the public generally.

APPENDIX.

CATALOGUE OF PLANTS WHICH GROW IN THE NEIGHBOURHOOD OF BUXTON, WITH A BOTANICAL COMMENTARY. BY MISS HAWKINS.

I DO not imagine that the following list is complete, as many plants may have been overlooked; and no mention is made of the grasses or mosses, owing to the difficulty of giving a correct list of them.

ELLEN HAWKINS.

ROCK HEAD, NEAR BUXTON,
4th May, 1854.

Proceeding along the road from Buxton to Bakewell, the first plant worthy of particular notice, is the Golden Saxifrage (Chrysoplenium oppositifolium). It grows very near the river, and flowers in May. Near the same place may be found the Red and Black Currant (Ribes rubrum, and R. nigrum), and also the Gooseberry (Ribes grossularia). But though it is natural for these shrubs to grow in damp soil, near a river, it is possible that the spot where these are found may be the remains of a former garden.

A little further, on the opposite side of the river, is a Willow, not very common,—the Sweet Bay-leaved Willow (Salix pentandra), producing its blossoms in June or July. Still further, on the other side of the river, and nearly opposite to the Lover's Leap, is a beautiful shrub, the Bird Cherry (Prunus padus), putting out its elegant spray of white blossoms in May. The ground beneath is covered with bright blue patches of the Great Water Scorpion-grass (Myosotis palustris), flowering from

June till August, and the White Wood Anemone (Anemone nemorosa), which flowers in April and May.

At the Lover's Leap, following the bed of the little stream which runs between high rocks, there will be found, in May and June, a great quantity of Broad-leaved Garlick (Allium ursinum), and at the upper end of the rocky bed of the rivulet, it is overhung by a beautiful deep pink Rose (Rosa villosa). This rose grows also in many places on the opposite bank of the river Wye, together with the Common Dog Rose (Rosa canina), and the White Trailing Dog Rose (Rosa arvensis).

In all the plantations, on each side of the road, may be found the Wild Raspberry (Rubus idæus).

Passing the Lover's Leap, on the rocks on the right hand, is a plant not found in any other place in England, except near Malham Cove in Yorkshire, the Blue Jacob's-ladder, or Greek Valerian (Polemonium cæruleum), flowering in June. Other plants on the side of the road are common everywhere. Smooth Speedwell (Veronica serpyllifolia), Germander Speedwell (Veronica chamædrys), Procumbent Speedwell (Veronica agrestis), Wall Speedwell (Veronica arvensis), Ivy-leaved Speedwell (Veronica hederifolia), all flowering from April till June, or even later. The Round-leaved Bell-flower (Campanula rotundifolia), flowering in July and August. The Field Scabious (Scabiosa arvensis), flowering in July. Small Scabious (Scabiosa columbaria), which is much more rare than the former, and flowers from June till August. Cuckoo-flower (Cardamine pratensis) flowers in April and May. Herb Robert, or Stinking Crane's-bill (Geranium Robertianum), Shining Crane's-bill (Geranium lucidum), Dove's-foot Crane's-bill (Geranium molle), all flowering from May to August. Near the river, the Butter-bur (Tussilago petasites) grows in profusion, producing larger leaves than any indigenous plant in Great Britain.

Beyond the mill, on the left hand of the road, the bridge leads to a bushy bank of small extent, skirting a foot-path leading to Fairfield; on which bank, besides the Prunus spinosa (Blackthorn, or Sloe), flowering in April or May, the Raspberry

(Rubus idæus), Common Blackberry (Rubus fruticosus), also flowering in April or May; the Dog's Violet (Viola canina), and Common Cowslip (Primula veris), and some rarer plants are to be found. Here are the Green Hellebore (Helleborus viridis), flowering in April or May; Clustered Bell-flower (Campanula glomerata), Bloody Crane's-bill (Geranium sanguineum), flowering from July to September; Hairy Violet (Viola hirta), flowering in April; Common Dwarf Cistus (Cistus helianthemum), and Hairy St. John's-wort (Hypericum hirsutum), both flowering in July and August. Bushes of a very beautiful deep pink or red Rose (Rosa tomentosa), also grow on this bank. Close to the river, a little farther on the opposite side, is the Common Dame's Violet (Hesperis matronalis), flowering in May and June.

Between the bridge and the turnpike, are found on the rocks, on the right hand—as also in many of the pastures—the Grass of Parnassus (Parnassia palustris), flowering in September and October; Ivy-leaved Wall Lettuce (Prenanthes muralis), flowering in July; Common Wall Cress (Arabis Thaliana), flowering in April; Hairy Wall Cress (Arabis hirsuta), flowering in May. The Nottingham Catchfly (Silene nutans) used to grow on this bank; but in taking materials for repairing the road, it seems to have been destroyed: it grows in the end of Millar's Dale, and in Dove Dale.

Beyond the turnpike, in the wet ground on the left hand, are the Common Marsh Marigold (Caltha palustris), flowering in March and April; Common Water Cress (Nasturtium officinale), flowering in June and July; and by the road side, chiefly on the inner side of the wall, is the Blue Meadow Crane's-bill (Geranium pratense), flowering in June and July. Still farther, past the bend of the road, are Small Marsh Valerian (Valeriana dioica), flowering in June; Wild Angelica (Angelica sylvestris), flowering in July; Great Hairy Willow-herb (Epilobium hirsutum), flowering in July. This last plant grows on the bank dividing the fish-pond from a small branch of the river, mixed with large patches of Blue Scorpion-grass, the Common Cow

Parsnep (Heracleum sphondylium), flowering in July. On the right hand of the road, before reaching this bend of the road, are the Small-flowered Hoary Willow-herb (Epilobium parviflorum), flowering in July ; Brook-lime (Veronica Beccabunga), Mouse-ear Chickweed - (Cerastium vulgatum), and Silver-weed, or Wild Tansy (Potentilla anserina), all flowering from May to July.

On approaching the third milestone from Buxton, the limestone is interrupted by toadstone, and as far as this extends, a beautiful little plant grows in the crevices of the stones, the Common Butterwort (Pinguicula vulgaris), flowering in May and June. In many parts of the river grows the White Floating Crowfoot (Ranunculus aquatilis), showing its white flowers in May. On the rocks on the right hand, going up the hill called Topley Pike, grows a sweet-scented plant, the Common Kidney Vetch, or Ladies'-finger (Anthyllis vulneraria), flowering in June and July ; and higher up on those rocks, the Stone Bramble (Rubus saxatile), flowering in June. In the valley below the road, on rocks close to the river, there are a few plants of the White Beam-tree (Pyrus Aria), flowering in May. The Giant Bell-flower (Campanula latifolia) grows near the road, going on to Ashford.

The plants to be found near Buxton, further from the public road, are the Common Privet (Ligustrum vulgare), growing on rocks in a dale near the bottom of Topley Pike, flowering in May and June ; where may also be found the Common Buckthorn (Rhamnus catharticus), flowering in May and June also. And in a dale leading to Chelmorton, the Burnet-leaved Rose (Rosa spinosissima) grows amongst the loose stones on the side of a hill.

Amongst the rocks and shrubs in other places, are two plants which used to be considered doubtful natives of Britain ; the Cinnamon Rose (Rosa cinnamomea), and the Mountain Globeflower (Trollius europæus), both flowering in May and June. Other more common plants are to be found in almost every meadow and road side. Such are the Common Corn Salad, or

Lamb's Lettuce (Fedia olitoria), flowering from April to June; Devil's-bit Scabious (Scabiosa succisa), flowering from August to October; Cross-wort Bed-straw (Galium cruciatum), flowering in May; Smooth Heath Bed-straw (Galium saxatile), flowering from June to August; Yellow Bed-straw (Galium verum), flowering in July and August; Cleavers, or Goose-grass (Galium Aparine), flowering all the summer; Greater Plantain (Plantago major), Hoary Plantain (Plantago media), having a very pleasant scent, and flowering from June to August; Great Burnet (Sanguisorba officinalis), flowering in June and July; Common Wall Pellitory (Parietaria offici-nalis), flowering from June till September; Common Ladies'-mantle (Alchemilla vulgaris), flowering from June to August; Field Ladies'-mantle, or Parsley Piert (Alchemilla arvensis), flowering from May to October; Procumbent Pearlwort (Sagina procumbens), flowering from May till October; Annual Small-flowered Pearlwort (Sagina apetala), flowering in May or June.

Common Primrose (Primula vulgaris), flowering in April; Autumnal Gentian (Gentiana amarella), flowering in August; Perennial Blue Flax (Linum perenne), flowering in June and July; Mill Mountain (Linum catharticum), flowering from June to August; Smooth Cow Parsley (Chærophyllum sylvestre), flowering in May; Sweet Cicely (Myrrhis odorata), flowering in May, not very common; Common Earth, or Pig-nut (Bunium flexuosum), flowering in May or June; Common Gout-weed (Ægopodium podagraria), flowering in June; Sheep's Sorrel (Rumex acetosella), flowering in June and July; Broad Smooth-leaved Willow-herb (Epilobium montanum), flowering in July; Bilberry, or Black Whortleberry (Vaccinium myrtillus), flower-ing in May; Common Ling (Calluna vulgaris), flowering in June and July; Cross-leaved Heath (Erica tetralix), flowering in July and August; Fine-leaved Heath (Erica cinerea), flowering from July to October; White Meadow Saxifrage (Saxifraga granulata), flowering in June; Rue-leaved Saxifrage (Saxifraga tridactylites), flowering in May; Mossy Saxifrage, or Ladies'-cushion (Saxifraga hypnoides), flowering in June; Common

Chickweed, or Stitchwort (Stellaria media), flowering all the year ; Vernal Sandwort (Arenaria verna), flowering from May to August ; Orpine, or Livelong (Sedum telephium), flowering in August ; Biting Stonecrop, or Wall Pepper (Sedum acre), flowering in June ; Common Wood Sorrel (Oxalis acetosella), flowering in April and May ; Corn Cockle (Agrostemma githago), flowering in June and July ; Meadow Lychnis, or Ragged Robin (Lychnis Flos cuculi), flowering in June ; Red Campion (Lychnis dioica), flowering in May and June ; Common Houseleek (Sempervivum tectorum), flowering in July ; Meadow-sweet (Spiræa ulmaria), flowering in June and July ; Spring Cinquefoil (Potentilla verna), flowering in April and May ; Strawberry-leaved Cinquefoil (Potentilla fragariastrum), flowering in April ; Water Avens (Geum rivale), flowering in June and July ; Dwarf Cistus (Cistus helianthemum), flowering in July and August ; Lesser Celandine (Ranunculus ficaria), flowering in April ; Bulbous Crowfoot (Ranunculus bulbosus), flowering in May ; Creeping Crowfoot (Ranunculus repens), flowering from June to August ; Common Bugle (Ajuga reptans), flowering in May ; Common Ground Ivy (Glechoma hederacea), flowering in April and May ; Red Dead-nettle, or Archangel (Lamium purpureum), flowering in May ; Common Marjoram (Origanum vulgare), flowering in July and August ; Wild Thyme (Thymus serpyllum), flowering in July and August ; Common Self-heal (Prunella vulgaris), flowering in July and August ; Common Yellow Rattle (Rhinanthus cristagalli), flowering in June ; Common Eye-bright (Euphrasia officinalis), flowering from June to September ; Knotty-rooted Figwort (Scrophularia nodosa), flowering in July ; Purple Foxglove (Digitalis purpurea), flowering in June and July ; Common Whitlow-grass (Draba verna), flowering in March and April ; Speedwell-leaved Whitlow-grass (Draba muralis), flowering in April and May ; this is considered rather a rare plant. Rock Hutchinsia (Hutchinsia petræa), flowering in April ; Common Shepherd's-purse (Thlaspi bursa pastoris), flowering all the summer ; Common Milkwort (Polygala vulgaris), flowers in

June and July, in a variety of colours—blue, white, and lilac. Common Furze, or Gorse (Ulex europæus), flowering in May; Common Bitter Vetch (Orobus tuberosus), flowering in May and June; Tufted Vetch (Vicia cracca), flowering in July and August; Lesser Yellow Trefoil (Trifolium minus), flowering in June and July; Common Bird's-foot Trefoil (Lotus corniculatus), flowering from June to September; Greater Bird's-foot Trefoil (Lotus major), flowering in July; Common Perforated St. John's-wort (Hypericum perforatum), and Imperforate St. John's-wort (Hypericum dubium), both flowering in July and August; Hairy St. John's-wort, (Hypericum hirsutum), flowering in June and July; Yellow Goat's-beard (Tragopogon pratensis), flowering in June; Corn Sow-thistle (Sonchus arvensis), flowering in August; Rough Hawk-bit (Apargia hispida), flowering in July; Common Mouse-ear Hawk-weed (Hieracium pilosella), flowering in June; Musk Thistle, (Carduus nutans), flowering in July and August; Spear Plume Thistle (Cnicus lanceolatus), flowering from June to September; Marsh Plume Thistle (Cnicus palustris), flowering in July and August; Meadow Plume Thistle (Cnicus pratensis), flowering in June; Woolly-headed Plume Thistle (Cnicus eriophorus), flowering in August; Colt's-foot (Tussilago farfara), flowering in March and April; Common Ragwort (Senecio Jacobæa), flowering in July and August; Moon Daisy (Chrysanthemum leucanthemum), flowering in July; Common Yarrow, or Milfoil (Achillea millefolium), flowering in July; Black Knapweed (Centaurea nigra), flowering from June to August; Greater Knapweed (Centaurea scabiosa), flowering in July and August; Common Salad Burnet (Poterium sanguisorba), flowering in July; Perennial Mercury (Mercurialis perennis), flowering in April and May; Rough Hawk's-beard (Crepis biennis), flowering in July; Spotted Cat's-ear (Hypochæris maculata), flowering in July.

On Axe Edge are found, besides the three common heaths (Calluna vulgaris, Erica tetralix, and Erica cinerea),—the Black Crowberry, or Crakeberry (Empetrum nigrum), a plant in appearance very like a heath, but of a very different class, as it bears

.the blossoms and the fruit on separate plants ; the former appearing in May, and the latter, which is a black berry, like a small black currant, ripe about August. It is said, that in former times a sort of wine was made from this fruit in Iceland and Norway. The Bilberry (Vaccinium myrtillus) grows there in abundance, and also the Cowberry, or Whortleberry (Vaccinium vitis idæa), flowering in June, bearing a red berry. Cranberry, or Marsh Whortleberry (Vaccinium oxycoccus), flowering in June ; the fruit of which is, in an early state, pale coloured with red spots—but when fully ripe of a deep red. The Mountain Bramble, or Cloudberry (Rubus chamæmorus), which in June bears one very elegant white flower, on a slender stalk ; the fruit resembles a small white raspberry. The Wild Rosemary, or Marsh Andromeda (Andromeda polifolia), flowering in June.

The plants of the Orchis tribe, near Buxton, are the Early Purple Orchis (Orchis mascula), flowering in April and May ; Spotted Palmate Orchis (Orchis maculata), flowering in June and July ; Aromatic Palmate Orchis (Orchis conopsea), flowering in June ; Frog Orchis (Orchis viridis), on the high ground above the stables at Buxton, and flowering in June and July ; Green Man Orchis (Aceras anthropophera), flowering in June ; Common Twayblade (Listera ovata), flowering in June.

Plants of the Fern tribe immediately near Buxton, are the Common Polypody (Polypodium vulgare) ; Rigid Three-branched Polypody (Polypodium calcareum) ; Male Shield Fern (Aspidium filix mas), Broad Sharp-toothed Shield Fern (Aspidium dilatatum) ; Female Shield Fern (Aspidium filix foemina) ; Brittle Bladder Fern (Cyathea fragilis) ; Common Maiden-hair Spleenwort (Asplenium trichomanes) ; Green Maiden-hair Spleenwort (Asplenium viride), much more rare than the Common Spleenwort ; Wall Rue Spleenwort (Asplenium ruta muraria) ; Common Hart's-tongue (Scolopendrium vulgare) ; Northern Hard Fern (Blechnum boreale) ; Common Ovate Adder's-tongue (Ophioglossum vulgatum). Common Moon-wort (Botrychium lunaria) is said to grow near Corbar wood.

The Common Brake (Pteris aquilina) grows in abundance near Ashford, and in the grounds at Chatsworth.

Classification of Plants, which grow in the Neighbourhood of Buxton ; and most of which grow within the district immediately around the Town :—

Diandria.—Monogynia.

Ligustrum vulgare	. .	Common Privet.
Fraxinus excelsior	. . .	Common Ash.
Veronica serpyllifolia	. .	Smooth Speedwell.
„ Beccabunga	. .	Brook-lime.
„ anagallis	. .	Water Speedwell.
„ officinalis	. . .	Common Speedwell.
„ chamædrys	. .	Germander Speedwell.
„ agrestis	. . .	Procumbent Field Speedwell.
„ arvensis	. . .	Wall Speedwell.
„ hederifolia	. .	Ivy-leaved Speedwell.
Pinguicula vulgaris	. .	Common Butterwort.

Triandria.—Monogynia.

Valeriana dioica	. . .	Marsh Valerian.
Fedia olitoria	. . . .	Common Lamb's Lettuce.

Tetrandria.—Monogynia.

Scabiosa succisa	. . .	Devil's-bit Scabious.
„ arvensis	. . .	Field Scabious.
„ columbaria	. . .	Small Scabious.
Galium cruciatum	. . .	Cross-wort Bedstraw.
„ saxatile	. . .	Smooth Heath Bedstraw.
„ pusillum	. . . .	Least Mountain Bedstraw.
„ verum	. . . .	Yellow Bedstraw.
„ Aparine	. . .	Goose-grass.
Plantago major	. . .	Greater Plantain.
„ media	. . .	Hoary Plantain.
Sanguisorba officinalis	. .	Great Burnet.
Cornus sanguinea	. . .	Wild Cornet-tree.
Parietaria officinalis	. .	Wall Pellitory.

| Alchemilla vulgaris | . . | Common Ladies'-mantle. |
| „ arvensis | . . . | Field Ladies'-mantle. |

Tetrandria.—Tetragynia.

| Sagina procumbens | . . | Procumbent Pearlwort. |
| „ apetala | . . . . | Small-flowered Pearlwort. |

Pentandria.—Monogynia.

Myosotis palustris .	.	Water Scorpion-grass.
„ sylvatica	. .	Upright Wood Scorpion-grass.
Cynoglossum officinale	.	Common Hound's-tongue.
Symphytum officinale	. .	Common Comfrey.
Primula veris	. .	Common Cowslip.
„ vulgaris	. .	Common Primrose.
Polemonium cæruleum	. .	Greek Valerian.
Campanula rotundifolia	. .	Round-leaved Bell-flower.
„ latifolia	. .	Giant Bell-flower.
„ Trachelium	. .	Nettle-leaved Bell-flower.
„ glomerata	. .	Clustered Bell-flower.
Viola hirta	. . .	Hairy Violet.
„ canina .	. .	Dog's Violet.
„ lutea	. . .	Yellow Mountain Violet.
Rhamnus catharticus	.	Common Buckthorn.
Ribes rubrum .	. . .	Common Currant.
„ alpinum	. .	Tasteless Mountain Currant.
„ nigrum	. . .	Black Currant.
„ grossularia	. .	Common Gooseberry.
Hedera helix	. . .	Common Ivy.

Pentandria.—Digynia.

Chenopodium Bonus Henricus	Mercury Goose-foot.
Gentiana amarella .	. .	Autumnal Gentian.
„ campestris .	. .	Field Gentian.
Chærophyllum sylvestre	.	Wild Chervil.
Myrrhis odorata	. .	Sweet Cicely.
Bunium flexuosum	.	Common Earth-nut.
Ægopodium podagraria	. .	Common Gout-weed.

Angelica sylvestris   . . Wild Angelica.
Heracleum sphondylium . . Common Cow Parsnep.
Pimpinella magna .   .   . Greater Burnet Saxifrage.

Pentandria.—Tetragynia.

Parnassia palustris . . . Common Grass of Parnassus.
Linum perenne   .   .   . Perennial Blue Flax.
   „   catharticum . . . Mill Mountain.

Hexandria.—Monogynia.

Allium ursinum   .   . . Broad-leaved Garlick.
   „   vineale   .   . . Crow Garlick.
Juncus lampocarpus . . Shining Jointed Rush.

Hexandria.—Trigynia.

Rumex sanguineus . . Bloody-veined Dock.
   „   acetosa   .   .   . Common Sorrel.
   „   acetosella . . Sheep's Sorrel.

Octandria.—Monogynia.

Epilobium hirsutum . . Hairy Willow-herb.
   „   angustifolium . Rose-bay Willow-herb.
   „   parviflorum . Small Willow-herb.
   „   montanum . Broad Smooth Willow-herb.
Vaccinium myrtillus . . Bilberry.
   „   vitis idæa . . Red Whortleberry.
   „   oxycoccus . Cranberry.
Calluna vulgaris . . Common Ling.
Erica tetralix .   .   . Cross-leaved Heath.
   „   cinerea .   .   . Fine-leaved Heath.

Octandria.—Tetragynia.

Paris quadrifolia . . Common Herb Paris.

Decandria.—Monogynia.

Andromeda polifolia . . Marsh Andromeda.

Decandria.—Digynia.

Chrysoplenium alternifolium . Alternate-leaved Golden Saxifrage.

„ oppositifolium Opposite-leaved Golden Saxifrage.

Saxifraga granulata . . Meadow Saxifrage.
„ tridactylites . . Rue-leaved Saxifrage.
„ hypnoides . . Mossy Saxifrage.

Decandria.—Trigynia.

Silene nutans . . . . Nottingham Catchfly.
Stellaria media . . . Common Chickweed.
Arenaria verna . . . . Vernal Sandwort.

Decandria.—Pentagynia.

Sedum Telephium . . . Orpine Livelong.
„ acre . . . . Biting Stonecrop.
Oxalis acetosella . . . . Common Wood-sorrel.
Lychnis flos cuculi . . . Meadow Lychnis.
„ dioica . . . Red or White Campion.
Cerastium vulgatum . . . Mouse-ear Chickweed.
Spergula nodosa . . . Knotted Spurrey.

Dodecandria.—Dodecagynia.

Sempervivum tectorum . . Houseleek.

Icosandria.—Monogynia.

Prunus padus . . . Bird-cherry.
„ spinosa . . . . Sloe Blackthorn.

Icosandria.—Di-Pentagynia.

Mespilus oxyacantha . . Hawthorn.
Pyrus aucuparia . . . Mountain Ash.
„ Aria . . . . White Beam-tree.
Spiræa filipendula . . . Common Dropwort.
„ ulmaria . . . Meadow-sweet.

Icosandria.—Polygynia.

Rosa cinnamomea . . .	Cinnamon Rose.
„ spinosissima . . .	Burnet Rose.
„ villosa	Soft-leaved Rose.
„ canina	Common Dog Rose.
„ arvensis	White-trailing Rose.
Rubus fruticosus . . .	Blackberry.
„ idæus	Raspberry.
„ corylifolius . . .	Hazel-leaved Bramble.
„ saxatilis	Stone Bramble.
„ chamæmorus . .	Cloudberry.
Potentilla anserina . . .	Silver-weed.
„ verna . . .	Spring Cinque-foil.
„ fragariastrum . .	Strawberry-leaved ditto.
Geum urbanum . . .	Common Avens.
„ rivale	Water Avens.

Polyandria.—Monogynia.

Cistus helianthemum . .	Common Cistus.

Polyandria.—Pentagynia.

Aquilegia vulgaris . . .	Columbine.

Polyandria.—Polygynia.

Anemone nemorosa . .	Wood Anemone.
Ranunculus ficaria . . .	Crowfoot.
„ bulbosus . .	Bulbous Crowfoot.
„ repens . . .	Creeping Crowfoot.
Trollius europæus . . .	Mountain Globe-flower.
Helleborus viridis . . .	Green Hellebore.
„ fœtidus . .	Stinking Hellebore.
Caltha palustris . . .	Marsh Marigold.

L

Didynamia.—Gymnospermia.

Ajuga reptans . . .	Common Bugle.
Glechoma hederacea . . .	Ground Ivy.
Lamium purpureum . .	Red Dead Nettle.
Origanum vulgare . . .	Common Marjoram.
Thymus serpyllum . .	Common Thyme.
„ acinos	Basil Thyme.
Prunella vulgaris . . .	Self-heal.

Didynamia.—Angiospermia.

Rhinanthus cristagalli . .	Yellow Rattle.
Euphrasia officinalis . .	Eyebright.
Scrophularia nodosa . . .	Knotty Figwort.
Digitalis purpurea . .	Purple Foxglove.

Tetradynamia.—Siliculosa.

Draba verna	Common Whitlow-grass.
„ incana . . .	Twisted Podded Whitlow-grass.
„ muralis	Speedwell-leaved Whitlow-grass
Hutchinsia petræa . .	Rock Hutchinsia.
Thlaspi bursa pastoris .	Shepherd's Purse.

Tetradynamia.—Siliquosa.

Cardamine impatiens .	Ladies' Smock.
„ pratensis . . .	Cuckoo-flower.
Nasturtium officinale . .	Water-cress.
Hesperis matronalis . . .	Dame's Violet.
Arabis thaliana . . .	Wall-cress.
„ hirsuta	Hairy Wall-cress.

Monodelphia.—Decandria.

Geranium pratense . .	Blue Crane's-bill.
„ Robertianum . .	Herb Robert.

Geranium lucidum   . . Shining Crane's-bill.

    „   molle . . Dove's-foot Crane's-bill.

    „   sanguineum . . Bloody Crane's-bill.

Diadelphia.—Octandria.

Polygala vulgaris   . . . Common Milkwort.

Diadelphia.—Decandria.

Ulex europæus   . . . Common Furze.

Anthyllis vulneraria . . . Kidney Vetch.

Orobus tuberosus . . . Common Bitter Vetch.

Vicia sylvatica . . . Wood Vetch.

    „   cracca . . . Tufted Vetch.

Hippocrepis comosa . . . Horse-shoe Vetch.

Trifolium minus   . . . Lesser Yellow Trefoil.

Lotus corniculatus   . . Bird's-foot Trefoil.

    „   major . . . . Greater Bird's-foot Trefoil.

Polydelphia.—Polyandria.

Hypericum perforatum   . . Perforated St. John's-wort.

    „   dubium . . Imperforated St. John's-wort.

    „   montanum . . Mountain St. John's-wort.

    „   hirsutum . . Hairy St. John's-wort.

Syngenesia.—Polygamia æqualis.

Tragopogon pratensis   . . Yellow Goat's-beard.

Picris hieracioides . . . Hawkweed Ox-tongue.

Sonchus arvensis   . . Corn Sow-thistle.

Prenanthes muralis   . . Ivy-leaved Wall Lettuce.

Leontodon taraxacum   . . Dandelion.

Apargia hispida   . . Rough Hawkbit.

Hieracium pilosella . . . Mouse-ear Hawkweed.

    „   murorum . . Broad-leaved Wall Hawkweed.

    „   umbellatum . . Narrow-leaved Hawkweed.

Serratula tinctoria   . . Common Saw-wort.

Carduus nutans . . . . Musk Thistle.
   „   acanthoides   . . Welted Thistle.
Cnicus lanceolatus   . . Spear Plume Thistle.
   „   palustris . . Marsh Plume Thistle.
   „   eriophorus . . Woolly-headed Plume Thistle.
   „   heterophyllus . . Melancholy Plume Thistle.
Carlina vulgaris   . . Carline Thistle.
Eupatorium cannabinum   . Hemp Agrimony.

Syngenesia.—Polygamia superflua.

Tussilago farfara   . . Colt'sfoot.
   „   petasites . . . Butter-bur.
Senecio vulgaris   . . Common Groundsel.
   „   Jacobæa . . Common Ragwort.
   „   tenuifolius . . Hoary Ragwort.
Solidago virgaurea   . . Common Golden Rod.
Bellis perennis . . . Common Daisy.
Chrysanthemum leucanthe-
mum . . . . } Ox-eye Daisy.
Achillea millifolium . . Milfoil.

Syngenesia.—Polygamia frustranea.

Centaurea nigra   . . Black Knapweed.
   „   scabiosa . . Greater Knapweed.

Gynandria.—Monandria.

Orchis bifolia   . . Butterfly Orchis.
   „   pyramidalis . . Pyramidal Orchis.
   „   mascula . . Early Purple Orchis.
   „   viridis . . Frog Orchis.
   „   maculata . . Spotted Palmate Orchis.
   „   conopsea . . Aromatic Palmate Orchis.
Aceras authropophera . Green Man Orchis.
Listera ovata   . . Common Twayblade.

Monœcia.—Tetrandria.

Urtica dioica . . . . Great Nettle.

Monœcia.—Pentandria.

Bryonia dioica . . . . Red-berried Bryony.

Monœcia.—Polyandria.

Arum maculatum . . . Cuckow Pint.
Poterium sanguisorba . . Salad Burnet.
Corylus Avellana . . . Common hazel-nut.

Diœcia.—Diandria.

Salix pentandra . . . . Bay-leaved Willow.
  " caprea . . . . Round-leaved Willow.
  " alba . . . . . Common White Willow.

Diœcia.—Triandria.

Empetrum nigrum . . Black Crowberry.

Diœcia.—Enneandria.

Mercurialis perennis . . . Perennial Mercury.

Diœcia.—Monadelphia.

Taxus baccata . . . Common Yew.

FERNS.
Cryptogamia.—Filices.

Polypodium vulgare . . . Common Polypody.
  " calcareum . . Rigid three-branched Polypody.
Aspidium Filix mas . . . Male Shield Fern.
  " aculeatum . . Prickly Shield Fern.
  " lobatum . . . Close-leaved Prickly Shield Fern.
  " dilatatum . . Sharp-toothed Shield Fern.
  " Filix fœmina . . Female Shield Fern.

Cystea fragilis . . .	Brittle Bladder Fern.
„ dentata . .	Toothed Bladder Fern.
Asplenium Trichomanes	Spleenwort.
„ viride . . .	Green Maiden-hair Spleenwort.
„ ruta muraria .	Wall-tree Spleenwort.
„ Adiantum nigrum	Black Maiden-hair Spleenwort.
Scolopendrium vulgare . .	Common Hart's-tongue.
Blechnum boreale . . .	Northern Hard Ferns.
Pterisaquilina . . .	Common Brakes.
Botrychium lunaria . . .	Moonwort.
Ophioglossum vulgatum .	Adder's-tongue.
Equisetum limosum . . .	Smooth Naked Horsetail.

DIRECTORY, &c.

DISTANCES—ROUTES—NEAREST RAILWAY STATIONS—COACHES—HORSES AND CARRIAGES FOR HIRE—LIST OF HOTELS, INNS, AND LODGING-HOUSES.

BUXTON is 159 miles from London, 38 miles from Derby, 22 miles from Matlock-Bath, 12 miles from Bakewell, 26 miles from Sheffield, 23 miles from Chesterfield, 15 miles from Chatsworth, 15 miles from Rowsley, 10 miles from Castleton, 6 miles from Chapel-en-le-Frith, 24 miles from Manchester, 17 miles from Stockport, 11 miles from Disley, 6 miles from Whaley, 12 miles from Macclesfield, 18 miles from Chelford, 12 miles from Leek, and 20 miles from Ashbourne. Rowsley, Stockport, Manchester, Macclesfield, Chelford, and Leek, are the nearest points at which the lines of railway facilitate the access to Buxton. The railway which terminates at Rowsley joins the Midland railway at Ambergate, having passed through Matlock and Darley Dale. This is generally found to be the most convenient route to Buxton from the south, south-west, and south-east, —as from London, Northampton, Leicester, Peterborough, Coventry, Birmingham, Worcester, &c. This is usually the more convenient route likewise from the Chesterfield, Sheffield, and Nottingham districts. From South Wales,

Shropshire, and the Potteries, the more convenient railway termini are usually Leek, Macclesfield, Chelford, or Stockport. From the north, the route from the east coast,—Newcastle-upon-Tyne, &c.,—is commonly by the Midland railway to Rowsley; and from the west coast,—Carlisle, Glasgow, &c.,—by Manchester, or Stockport. It is strange that, in these days of enterprise, and constantly increasing facilities of communication, the great Buxton district should not have been traversed long ago by a railway applicable to passenger-traffic; and 'this is the more strange, as the shortest route between Manchester and London would pass through Buxton; and as, although the elevation of Buxton is so considerable, the valleys are said to offer great engineering facilities, with moderately easy gradients; and, as it has been stated, that no very great tunnelling difficulties would have to be encountered. In the meantime, however, only an old-fashioned luggage railway, encumbered by many steep inclined planes, and therefore not adapted for the continuous use of locomotive engines, is at a less distance than twelve miles from Buxton.

There is daily communication, by means of coaches, between this place and Manchester, and Derby, at all times of the year. A mail-coach leaves Derby at eight o'clock every morning, and passes through Buxton, for Manchester, at half-past one o'clock; and a corresponding coach leaves Manchester every forenoon at ten o'clock, and passes through Buxton, for Derby, at half-past one o'clock. One or more coaches leave Buxton for Manchester every morning throughout the year, at half-past eight o'clock, and leave Manchester for Buxton at one o'clock p.m. From May or June to September or October, the coaches are multiplied in such numbers—starting from and returning to Buxton at various hours—that Manchester may be reached

from Buxton, or left for Buxton, at most times of the
day. During these months, there are likewise coaches to
and from Sheffield, Rowsley, and Macclesfield, every
day; and daily omnibuses to and from Chatsworth and
Haddon.

As to post-horses, and close and open carriages, drawn by
one or two horses, mule-carriages, donkey-carriages, and
Bath-chairs, there is every variety for hire, by the day, hour,
or mile, according to wish.

HOTELS, INNS, AND LODGING HOUSES.

LOWER BUXTON.

HOTELS AND INNS.

St. Anne's Hotel, the Crescent.—Mrs. Harrison.
The Old Hall, at the west end of the Crescent.—Mr. Bates.
The George, opposite the Square.—Mr. Lees.
The Grove, opposite the Hot Baths.—Miss Wood.
The Shakespeare, Spring Gardens.—Miss Barlow.
The White Lion, Spring Gardens.—Mr. Sutton.

LODGING HOUSES.

THE CRESCENT.

Great Hotel Board and Lodging House, No. 1.—Mr. Hicklin.
   „   „   „   „ 2.—(Post-office) Mr. Smilter.
   „   „   „   „ 3.—Miss Gregory.

THE SQUARE.

No. 1. Miss Glazbrook.   No. 4. Mr. Broomhead.
 „ 3. Miss Muirhead,   „ 6. Mrs. Moore.

HALL BANK.

*Leading to Upper Buxton, on the west of the Terrace Walks, and overlooking
these walks, the Crescent, and Fairfield.*

No. 1. Mr. James Turner.   No. 5. Mr. John Clayton.
 „ 2. Miss Clayton.   „ 6. Mrs. Robert Bates.
 „ 3. Mr. Joseph Hoyle.   „ 8. Mr. Millar.
 „ 4. Mr. Joseph Clayton.

QUADRANT,

Opposite the Hot Baths.

No. 1. Mr. Samuel Turner.

LOWER BUXTON—(CONTINUED.)

SPRING GARDENS,

Street leading from the Crescent to the Bakewell and Fairfield Roads.

Miss Flint.	Mr. Obadiah Hoyle.	Mr. John Clayton.
Mr. Smith.	Mr. Barrow.	Mr. Orgill.
Mr. Gregory.	Mr. Holmes.	Mrs. Grant.
Mrs. Francis.	Mrs. Cox.	Mr. T. Marshall.
Mr. Dineley.	Mrs. Swann.	Mr. Gibbons.
Mr. T. Webster.	Mr. Turner.	Mr. Widowson.
Mr. Sutton.	Mrs. Pidcock.	Mr. Barnsley.
Mr. William Evans.	Miss Illingworth.	Mrs. Martin.
Mr. Wainwright.	Miss Anzani.	Mr. Mellor.
Miss Sanders.	Mr. John Norton.	Mr. Johnson.
Mrs. Vickers.	Mr. R. Gregory.	Mr. Bradshaw.

TERRACE ROAD, OR YEOMAN'S LANE,

On the east of Terrace Walks; leading to Upper Buxton.

Mrs. Street.	Mr. A. Plant.	Miss Clayton.
Mr. Chapman.	Mr. Oldfield.	Mr. Bailey.
Miss Jeffcote.	Mr. G. Smith.	Mr. Bradbury.
Mr. John Clayton.		

UPPER BUXTON.

HOTELS AND INNS.

The Eagle.—Mr. Wood.
King's Head.—Mr. G. Brown.
New Inn.—Mr. Hartshorn.
Queen's Head.—Mr. G. Hobson.
Seven Stars.—Mr. Golland.
Sun.—Mr. Stubbs.
Cheshire Cheese.—Mr. J. Brown.
Swan.—Mr. B. Mycock.

LODGING HOUSES.

Mr. Rowland.	Mr. Fidler.	Mr. Hulse.
Mr. R. Deakin.	Mrs. Thompson.	Mr. Cantrell.
Mrs. Bennett.	Mr. T. Ingham.	Mr. Goddard.
Mr. H. Clough.	Mr. D. Wheeldon.	Mr. Simpson.
Mr. James Clayton.	Mr. J. Boam.	Mr. Morton.
Mr. E. Hobson.	Mr. F. Evans.	Mr. Debiolel.
Mr. T. Walton.	Mr. Ward.	Mr. Spooner.
Miss Deakin.	Mr. Brocklehurst.	Mr. Brunt.
Mr. G. Smith.	Miss Poulson.	Mr. H. Morton.

UPPER BUXTON—(CONTINUED.)

Mr. Henshaw.
Mr. Whalley.
Mr. Percival.
Mr. Isaac Brunt.
Mr. Ball.
Mr. Worrall.
Mr. Woodruff.
Mrs. Perkin.
Mr. E. Webster.
Mr. Swinscow.
Mr. Street.
Mrs. Fidler.
Mr. Tyson.
Mr. T. Lees.

Mrs. Clayton.
Mr. J. Robinson.
Mr. Downs.
Miss Percival.
Mr. Marshall.
Mr. Chapman.
Mrs. Drabble.
Mr. John Turner.
Mr. T. Brunt.
Mr. Wildgoose.
Mr. Perkin.
Mr. Kitchen.
Mr. Brandreth.
Miss Sharland.

Mr. Brown.
Miss Ward.
Mr. R. Smith.
Mr. Norton.
Mrs. Robinson.
Mr. W. Deakin.
Mr. W. Ward.
Mr. Redfern.
Mr. Baguley.
Mrs. Fidler.
Mr. George Taylor.
Mr. Kitchen.
Mr. Newham.
Mr. J. Wild.

THE END.

Also, in two volumes, post 8vo, cloth, pp. 355 and 353, price 12s.

A TREATISE ON DIET AND REGIMEN.

FOURTH EDITION.

RE-WRITTEN AND MUCH ENLARGED.

EMBRACING THE MORE RECENT VIEWS, FACTS, AND DISCOVERIES OF CHEMISTRY AND STATISTICS.

"The first edition of this work appeared so long back as 1835. The treatise was short, condensed, sensible, and perspicuous; and contained a view of dietetics in which, with close adherence to scientific principles, information on all topics relating to dietetic and hygienic management was conveyed in a popular and interesting form. Since that time the work has passed through three editions, in each of which the author has studied to introduce all those improvements which the gradual progress of science during the interval, and his own experience in a situation where he has occasion to observe much of the effects of diet, have enabled him to do.

"The present is in all respects a greatly improved form of the work. It is manifest that Dr. Robertson has been at no ordinary pains in explaining the scientific principles of diet, according to the most recent and most approved chemical and physiological doctrines. . . . In conclusion we have only to say that, while the medical reader seeking information will peruse this volume with advantage, it must prove highly useful to the general reader and invalid, or dyspeptic, in warning what to avoid and what to choose."—*Edinburgh Medical and Surgical Journal*.

"A sensible and useful book. . . . It is written in a plain unaffected manner, and contains much valuable information."—*British and Foreign Medical Review*.

"A good work."—*Medico-Chirurgical Review*.

"It is scarcely necessary that we should add our hearty recommendation of Dr. Robertson's treatise, not merely to our medical readers, but to the public over whom they have an influence. It is one of the few books which is legitimately adapted, both in subject and manner of treatment, to both classes."—*British and Foreign Medico-Chirurgical Review*.

"We noticed the appearance of the first volume of this work, on a former occasion, with approval. We have since read the second volume, and find reason to repeat our approbation. . . . A work which is creditable to his judgment, and calculated to prove serviceable to his readers; we therefore feel no hesitation in recommending these volumes to every one desirous of accumulating a choice medical library."—*Dublin Quarterly Journal of Medical Science*.

"The author has displayed considerable industry and research, affording much information and many useful directions. . . . The treatise will form a valuable addition to our stock of knowledge on the subject of which it treats, brought up, as it is, to the present state of chemical science."—*Lancet*.

"As it has been wholly re-written, and the view given on the subject brought up with the current of science, it may be regarded as a new work. . . . It deserves to be characterised as a very sensible book; plainly the production of a man of much intelligence, of extensive reading, and of large experience in his profession. . . . Altogether, the work is worth the attention of the public, as well as of the profession."—*Monthly Journal of Medical Science*.

"This valuable work, by the physician of the Buxton Bath Charity, is now in its fourth edition. Well worthy the attention of parents and guardians, as well for their own sakes, as for the health and happiness of those entrusted to their care."—*Morning Post*.

"The great recommendation, in our eyes, to this work is, that there is no quackery in it—no over-doing you with wisdom—no panaceas. Its language is that of a friend speaking to a friend, in plain, intelligible, modest, but impressive language, on subjects of all but the highest importance to every one who feels that he is an intelligent, and acknowledges himself to be an accountable, moral agent."—*Nottingham Mercury*.

"This book,—which it would not be too much to call a supplementary volume to the Reports of the Sanatory Commissioners, and to characterise as being fully as necessary to be studied ere the social condition of the people can be rightly understood, or efficiently rectified, – ought to be read attentively by all who would deal beneficially with the health of the country."—*Preston Chronicle.*

"It would be found an admirable adjunct to the labours of the Executive, in promoting an improvement in the sanatory condition of the empire."—*Bath Herald.*

"Wise and weighty. . . . Demands the attention of the 'Powers that be.' . . . Dr. Robertson has proved that a wise man can make even a deeply abstruse scientific subject perfectly readable."—*Derby Reporter.*

"Dr. Robertson's Treatise on Diet and Regimen is unequalled in the language." —*Sun.*

"We must conclude our brief notice of Dr. Robertson's work, and we do so by strongly recommending it to the reading public. Its matter is valuable, and in many respects original. It is a work in which common sense, aided by experience, is applied to subjects of the greatest consequence to our well-being, and to subjects which are seldom treated except in a narrow and bigoted spirit. The author is an observer and thinker for himself, and we may say for him—

> 'Nullius addictus jurare in verba magistri
> —— huic erit mihi magnus Apollo.' "

Constitutional Magazine.

Also, 12mo, cloth, price 2s. 6d.

BUXTON AND ITS WATERS:

AN ANALYTICAL ACCOUNT OF THEIR MEDICINAL PROPERTIES AND GENERAL EFFECTS.

"An interesting and judicious performance."—*Edinburgh Medical and Surgical Journal.*

"Will prove useful to the general as well as the professional reader."—*Medico-Chirurgical Review.*

Also, in a Cover, price 6d.

A GUIDE TO THE USE OF THE BUXTON WATERS.

EIGHTH EDITION, REVISED.

"The recent work of Dr. Robertson puts the reader in possession of the cases most benefited by these waters, better than I can do, from his long experience."— *Dr. Seymour on Several Severe Diseases of the Human Body.*

"To those who think to pay Buxton a visit, or may propose to send patients there, we can recommend this little pamphlet, as giving very useful information respecting the use and curative effects of its tepid waters."—*Lancet.*

"We have derived much satisfaction from the perusal of this little pamphlet, which will be found a necessary companion to all who are inclined to pay a visit to the Buxton Springs."—*Medical Gazette.*

Also, price One Shilling.

A LETTER

TO

DR. LYON PLAYFAIR, C.B., F.R.S.:

BEING

A MEDICAL COMMENTARY ON THE RESULTS OF THE RECENT
ANALYSIS OF THE BUXTON TEPID WATERS;

TO WHICH ARE PREFIXED A STATEMENT OF THE IMPROVEMENTS NOW IN
PROGRESS AT BUXTON,

AND DR. PLAYFAIR'S ANALYTICAL REPORT.

"Dr. Robertson is well and favourably known to the profession by his previous writings; and he is fortunate, on the present occasion, in dealing with a subject at once novel and important. It appears that Buxton has had, of late years, such an increasing influx of invalids, that the town and baths could not properly accommodate them, and the owner of the baths and adjacent property, the Duke of Devonshire, has consequently made great improvements to meet the public wants, under the superintendence of Sir Joseph Paxton. In connection with these alterations, which have quite transformed the facilities for bathing, the Duke requested Dr. Playfair to make an analysis of the thermal springs, which, it will be seen, has resulted in a discovery of great consequence to the reputation of these celebrated waters. It was found that every imperial gallon of the waters contained 206 *cubic inches of nitrogen;* and upon this gaseous element, Dr. Playfair considers the medicinal properties of the Buxton waters entirely to depend. We may judge of the importance of this discovery by the circumstance stated by Dr. Robertson, that hitherto only 5·57 cubic inches per gallon were supposed to be the proportion in which this important element was contained in these waters. As 120 gallons are discharged from the springs per minute, the whole discharge of nitrogen per minute amounts to 24,720 cubic inches.

"Dr. Robertson, after lucidly comparing the analysis of Dr. Playfair with other analyses, and with the analyses of other thermal waters, looks forward to new discoveries, and thus concludes his interesting pamphlet:—

"'I would venture to reiterate, as a possibly needful corollary to the above statements and inferences, the opinion, that all the medicinal effects of the Buxton tepid water, and especially its great alterative action, can scarcely be ascribed even to the large proportion of nitrogen which it must now be held to contain (significant and valuable as this must be admitted to be), but may still be referable to the presence of some hitherto undetected constituent. However this may be, the great and singularly lasting effects of these baths, and of the internal use of this water, are unquestioned and indisputable. Their use is almost specific for the relief or alleviation of most cases of rheumatism, and of many cases of gout, for which the use of other means and appliances has been sought and tried in vain. In proof of this, the fact may be adduced that large numbers of poor handicrafts-men, who have proved the effect of this water on their suffering and imperfectly usable limbs, are known to undergo great privations in order to secure its use at stated intervals, from finding that no other means within their reach enable them to maintain such a state of their joints as is needful to enable them to follow their employment. The yearly reports of the Buxton Bath Charity certify, that of 15,497 patients, for the most part sufferers from rheumatism, admitted to the benefit of the institution, from the year 1838 to 1851, only 613 had to be sent home as being "no better," the large proportion of 11,740 having been discharged as cured or much "relieved!"'

"These are great results, and may vie with any of the vaunts of the thermalists of Germany. They should attract our invalids to this beautiful locality, in preference to the continental watering places."—*The Lancet, Oct. 2nd,* 1852.

Charles M. Tayntor,
October, 21st, / A.D. 1884,
Manchester
Connecticut,

www.ingramcontent.com/pod-product-compliance
Lightning Source LLC
LaVergne TN
LVHW012204040326
832903LV00003B/105